THE FACE THAT DEMONSTRATES THE FARCE OF EVOLUTION

THE FACE THAT DEMONSTRATES THE FARCE OF EVOLUTION

HANK HANEGRAAFF

NASHVILLE

A Thomas Nelson Company

Published by Word Publishing
Nashville, TN

Library of Congress Cataloging-in-Publication Data

Hanegraaff, Hank.
The face that demonstrates the farce of evolution/by Hank Hanegraaff.
p. cm.
Includes bibliographical references and indexes.
ISBN 0-8499-4272-1 (trade paper)
ISBN 0-8499-1181-8 (hardcover)
1. Evolution (Biology)—Religious aspects—Christianity.
2. Darwin, Charles, 1809-1882. I. Title.
BT712.H36 1998
231.7'652—dc21

98-37940
CIP

Printed in the United States of America
1 2 3 4 5 PHX 9 8 7 6 5 4 3 2 1

To my wife, Kathy, "a woman of virtuous character, more valuable than jewels . . . her children rise up and call her blessed; her husband also."

The Face That Demonstrates the Farce of Evolution

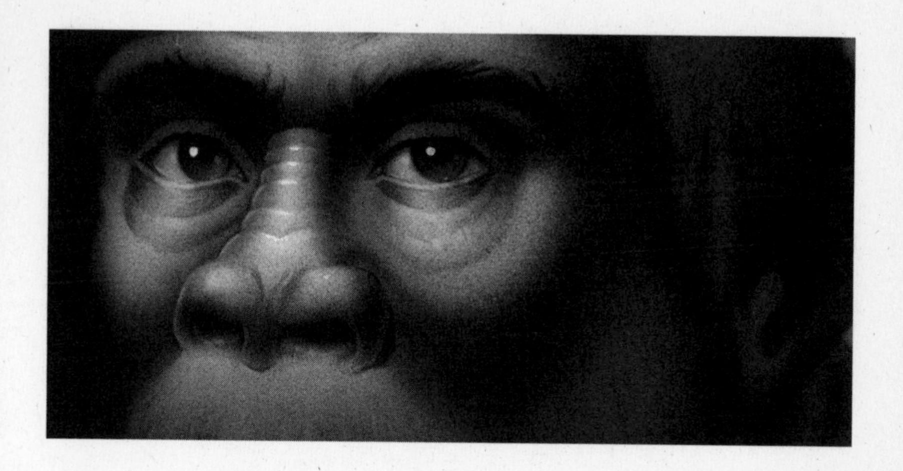

Contents

Foreword

Hank Hanegraaff has earned a reputation as a superlative Bible teacher and as a vibrant radio presence to whom many thousands of listeners turn when they need answers to hard questions. The subject of evolution needs as much good teaching as it can get. As one who has plowed that bit of field myself, I'm delighted to see Hank bringing his very special skills to the task of showing Christians what is at stake in the evolution controversy, and telling them why they can't afford to ignore it.

Something very strange happened to our society in the twentieth century. It wasn't just that the details of the Genesis account had to be reinterpreted to take account of scientific findings. No, the Darwinian revolution went much farther than that. The fundamental biblical understanding of reality, which tells us that an all-powerful and loving God created men and women for a purpose, lost ground first in the universities, and then in the culture at large. The spiritual cause of this decline was the perennial wish of fallen men to refuse to honor their true Creator, and to substitute idols or natural forces for the glory of the immortal God. This we know from Romans 1:20–23, and the inevitable moral decline that stems from that foolish substitution is described in the verses that follow. If humans are "really" just animals living by chance in an amoral universe, what else would you expect?

The Darwinian theory of evolution made it appear that "science" had validated itself. After the triumph of Darwinism, it seemed absurd to most intellectuals to suppose anything else. The term *agnostic* was invented by Darwin's first disciple, Thomas Huxley, to describe the indifference toward the idea of God that Darwinism fostered. Agnosticism supposes that we can have no knowledge of God, and hence we can safely ignore the whole subject. The Darwinists did not (and generally do not) claim to be able to prove that God does not exist, but they do claim to be able to account for the entire history of life without allotting any role to a creator.

That is why the Darwinian theory is so important, and so misleading. It tells us that a blind and purposeless material process is our true creator, and that "God" is merely a product of the primitive human imagination. Just as a child ceases to believe in Santa Claus when he grows old enough to know who really puts the presents into the stockings, the student of evolution is expected to recognize that God is a fantasy when he learns that matter is all there is and that matter had to do its own creating.

Most people who have read this far will understand that evolutionary naturalism is the false teaching of persons who "professing to be wise . . . became fools." It is not enough to know that this doctrine is wrong, however. We have to understand why such a shaky theory has so much power over the minds of highly educated persons, so that what is mainly philosophy is taught as fact—even in many Christian classrooms. Too many Christian parents have unwisely tried to protect their children from learning about evolution. Many of these parents have lost that battle when their children went off to college (or even high school) where they were defenseless against an apparently irresistible philosophical system falsely presented in the name of "science." What we need to do is not to shelter our young people, but to inoculate them by teaching them a great deal more about evolution than the mainstream science educators want them to know.

Hank Hanegraaff doesn't advise parents and teachers to hide from false teaching. He gives them memorable teaching devices to help them identify the fallacies and to help them teach young people not to be misled. He exposes the specific wrong answers and provides lots of references to other literature. He teaches Christians to avoid the bad arguments by which they sometimes discredit themselves and warns them to avoid some of the unreliable information that has been widely circulated by well-meaning Christians who don't have their facts right. Finally, he encourages us to stay with the big picture and not to go off on matters of detail that divide the faithful and give ammunition to the agnostics.

So Hank has done his many admirers a great service by providing this very helpful book. I hope that those who read it and learn from it will help us to create a new generation of educated Christians who stand up for good science, but who also know how to expose the bad thinking and hidden philosophical assumptions so pervasive in evolutionary theory.

Phillip E. Johnson
Jefferson E. Peyser Professor of Law
University of California, Berkeley

Acknowledgments

As with the world, this work did not come into being as a result of blind evolutionary processes. Rather, it was created through years of research and reflection. As the notes and bibliography demonstrate, I am deeply indebted to the input and insight of hundreds of authors and resources.

I am also deeply indebted to a host of others. Among them are Kathy and the kids—Michelle, Katie, David, John Mark, Hank, Jr., Christina, Paul Steven, and Faith. Once again they demonstrated remarkable patience during the research and writing process.

I am also grateful for the wonderfully cohesive board and staff that God has allowed us to assemble at the Christian Research Institute. I am especially thankful for the editorial input of Stephen Ross and for his ferocious commitment to detail and sound argumentation. His competency and Christlike character bear eloquent testimony to the reality of the Creator.

Above all, I am grateful to my heavenly Father for giving me the commitment and creativity necessary to complete this task. To Him be the glory!

The heavens declare the glory of God; the skies proclaim the work of his hands. Day after day they pour forth speech; night after night they display knowledge. There is no speech or language where their voice is not heard. Their voice goes out into all the earth, their words to the ends of the world.

—King David

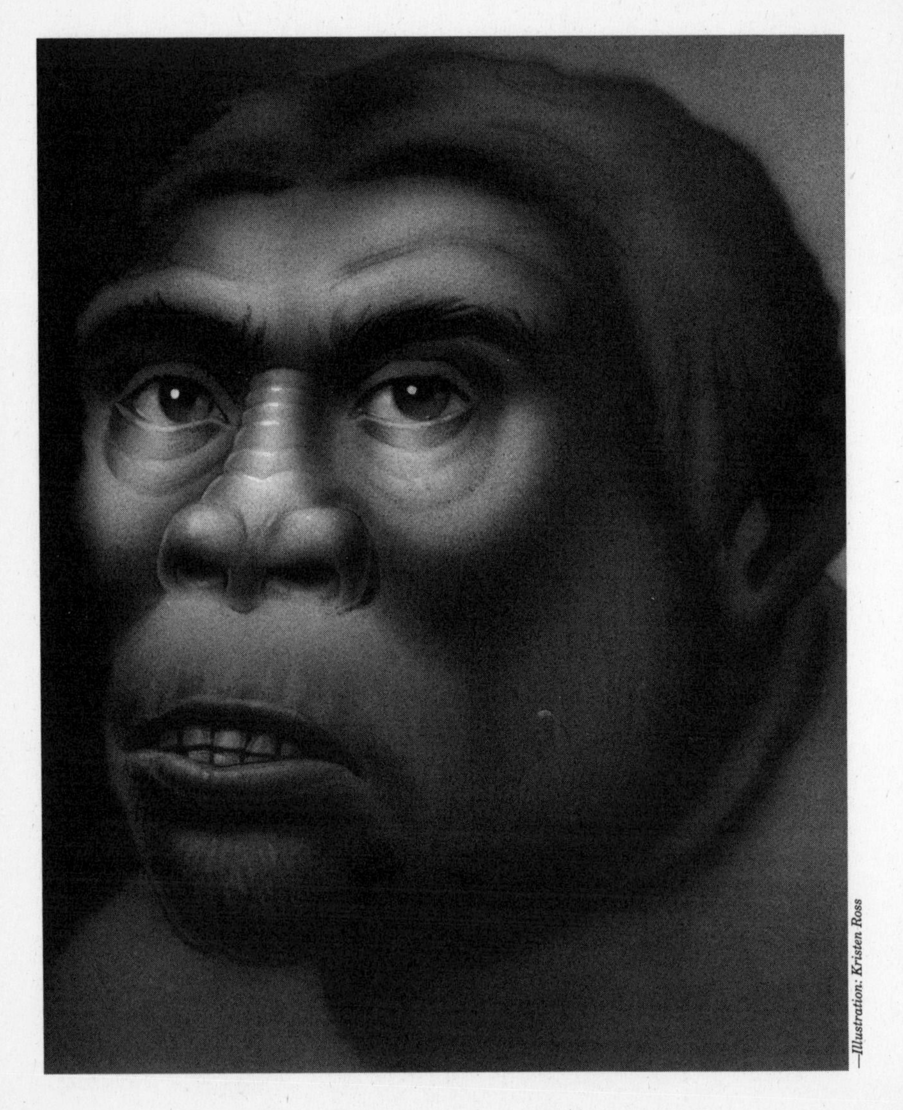

—Illustration: Kristen Ross

The *FACE* that demonstrates the farce of evolution.

Before You Begin

I'll never forget the day I first saw the *face* that demonstrates the *farce* of evolution. It virtually leapt off of the page of my textbook. It had a sharply receding forehead that cascaded abruptly into a heavy brow ridge. Its mouth jutted open, revealing apelike teeth. Its eyes were deep set and pensive; they were the eyes of a philosopher, with a slightly worried look as though he'd just seen his tax accountant.[1]

If ever a picture was worth a thousand words, this was it. Even at a tender age, I understood exactly what that picture meant. The message was crystal clear: *This monkey was man in the making!* My earliest ancestor was not Adam but an ape. The Bible was based on fiction. "In the beginning God created . . ." was simply the start of an infamous fairy tale.

Almost twenty years passed before I realized I'd been duped. The *face* that had stared back at me from the pages of my textbook was a *farce. Pithecanthropus erectus* (the ape-man who walked erect)—*not* Genesis—was the figment of a fertile imagination.[2]

From Evolution to Evidence

I began life's journey in the security of a Christian home, the product of a godly heritage. Each mealtime my father read Scripture, and each week I attended an orthodox church. My parents and my pastors taught me about a God who loves me with an everlasting love, about a God who is also righteous and just—whose "eyes are too pure to look on evil; [and who] cannot tolerate wrong" (Hab. 1:13; Jer. 31:3;). They made it clear that only through the person and work of Jesus Christ would my life have true meaning, purpose, and fulfillment.

But there was a catch. To experience true peace, I had to be willing to submit my life to Christ rather than travel the road of life guided by the dictates of my own will. And this I was not willing to do. I did not want to be deprived of all the pleasures the world had to offer, so I chose *rebellion* instead of *repentance.* It was not hard for me to justify my decision. At the same time that I was reading Scripture in church, I was reading science in class. And it was in science that I thought I had discovered an avenue of escape from accountability.

The prevailing scientific sentiment was that the world had come into existence through purely natural and random processes. Humans arose from the primordial slime: Atoms evolved into amoebas, invertebrates became vertebrates, and monkeys evolved into humans. "Enlightened" teachers were quick to point out that science had proved the Bible to be the stuff of legends and God to be little more than a cosmic kill-joy.

With the Creator out of the picture, I was free to grab for all the gusto life had to offer. And so, until age twenty-nine,

I followed the dictates of my own will, searching for happiness as I moved from one "happening" to the next. Despite temporary successes, I was left with a nagging emptiness that never seemed to go away.

One wet and windy January evening in 1979, however, three Christians paid me a visit. They reminded me that God loved me and had a transcendent purpose for my life. They claimed Jesus Christ could replace my emptiness with peace and the Bible was a road map that would guide me through the maze of life. They were so kind and thoughtful that I felt a tinge of guilt for responding to their compassion with sarcasm and condescension.

I could not resist the urge to point out that science had long ago demonstrated that God was little more than an illusion, that the Bible was a book for narrow-minded obscurantists, and that humans and their religious notions were merely functions of evolutionary processes.

My guests listened politely until I paused to catch my breath. Then one of them asked if I was open-minded enough to objectively consider the evidence on both sides of the origins debate so I could make an *informed* decision. They told me that the following week their church would be hosting a workshop in which evolution would be examined in the light of the empirical laws of science. With that, they left.

That next week, I was irresistibly drawn to a day of destiny—a day that would change my life for time and for eternity. As the day progressed, the foundation of evolution began to crumble and the shaky pillars on which I had built my worldview collapsed. More than that, my smug stereotyping of Christians as Bible thumpers who had checked their brains at the narthex of a church fell apart.

I came face to face with the realization that I had not rejected the creation model because the evidence for evolution was so overwhelming but because I did not want to yield my life to the will of the Creator. Not long after I examined the evidence, the Creator of the cosmos became the Lord and Savior of my life. Now, two decades later, He is more real to me than the very flesh upon my bones.

Today my passion is to demonstrate that evolution is—as Dr. Louis Bounoure, Director of Research at the French National Center for Scientific Research, put it—"a fairy tale for grown-ups."[3] More than that, it is my desire to equip committed Christians to use the fallacies of evolution as an opportunity to revolutionize the lives of multitudes for time and for eternity.

Whilst on board the Beagle I was quite orthodox, and I remember being heartily laughed at by several of the officers (though themselves orthodox) for quoting the Bible as a unanswerable authority on some point of morality.

—Charles Darwin,
The Life and Letters of Charles Darwin

When it was first said that the sun stood still and the world turned round, the common sense of mankind declared the doctrine false; but the old saying of Vox populi, vox Dei [voice of the people, voice of God] as every philosopher knows, cannot be trusted in science.

—Charles Darwin,
The Origin of Species by Means of Natural Selection

—Video: Courtesy of Eden Communications

Her Majesty's ship, the *Beagle*.

Charting the Course

On December 27, 1831, Charles Darwin left Devonport, England, aboard Her Majesty's Ship, the *Beagle.* Long before the departure date, however, Captain Robert FitzRoy began charting the course.

As Darwin set sail, he was a Bible-believing creationist. In his words, "I did not then in the least doubt the strict and literal truth of every word in the Bible. . . ."[1] However, as Dr. Michael Denton points out:

> For Darwin the *Beagle* proved the turning point of his life, a liberating journey through time and space which freed him from the constraining influence of Genesis. Every voyage conjures up a vision of new horizons and emancipation, but there is something particularly evocative about the voyage on the *Beagle* to the remote and little known shores of South America. It is almost as if the elemental forces of nature, so apparent along those cold and stormy

> coasts of Patagonia and Tierra del Fuego, had conspired together to fragment the whole framework of biblical literalism in Darwin's mind, to blow his intellect clear of all the accumulated cobwebs of tradition and religious obscurantism. The *Beagle* is also symbolic of the much greater voyage which the whole of our culture subsequently made from the narrow fundamentalism of the Victorian era to the skepticism and uncertainty of the twentieth century. Darwin's experiences during those liberating five years became the experience of the world.[2]

Denton goes on to demonstrate that what Darwin saw as the liberating worldview of evolution has in reality led civilization into a monstrous lie. Darwin's journey on the *Beagle* sowed seeds of doubt in his mind regarding creationism. As you continue through the pages of this book, you will discover that his doubts were totally unwarranted.

When Darwin charted his course, he did not know where his journey would lead. In contrast, as you travel through these pages, I want you to know exactly where you are going and how you will get there. If you begin as a creationist, you will become equipped to demonstrate the farce of evolution. In addition, your confidence in the Christian worldview will be solidified. If you begin as an evolutionist, the shaky pillars of your worldview will be undermined.

Truth or Consequences

We begin our journey with a chapter titled "Truth or Consequences." As you progress through this section, you will discover that more consequences for society hinge on the issue of human origins than on any other. Among them are the *sovereignty of self*, the *sexual revolution*, and *survival of the fittest*.

To make the heart of this material memorable, I developed chapters 2 through 5 around the acronym F-A-C-E. For generations, the "face" of *Pithecanthropus erectus* has been used covertly to communicate the notion that evolution is based in fact. I will now use it overtly to demonstrate that evolution is a farce.

Fossil Follies

The *F* in FACE will serve to remind us that the fossil record is an embarrassment to evolutionists. It demonstrates in spades that transitional forms from one species to another are purely mythological. While the general public seems blithely unaware of the fact that transitions from one species to another do not exist, it is common knowledge among paleontologists. That is precisely why novel theories involving *"pseudosaurs," pro-avises,* and *punctuated equilibrium* are constantly evolving.

Ape-Men Fiction, Fraud, and Fantasy

A represents Ape-Men Fiction, Fraud, and Fantasy. As you proceed, you will discover that *Pithecanthropus erectus* is fictitious; *Piltdown man* was a fraud; and *Peking man* is pure fantasy. To say that "hominids" like Peking man and his partners are closely related to humans because both can walk is like saying that a hummingbird and a helicopter are closely related because both can fly. The distance between an ape who cannot read or write and a descendant of Adam who can compose a musical masterpiece or send someone to the moon is the distance of infinity.

Chance

C stands for Chance. Imagine having the temerity to suggest that Christopher Wren had nothing to do with the design of St. Paul's Cathedral in London. Imagine asserting that the majestic *Messiah* composed itself apart from Handel. Or imagine that the *Last Supper* painted itself without Leonardo da Vinci. In chapter 4 we will consider some even more egregious notions—that an *eye,* an *egg,* or the *earth,* each in its vast complexity, came into existence by blind chance. As we will see, forming even a protein molecule by random processes is not only improbable but is also, indeed, impossible.

Empirical Science

E represents Empirical Science. Go to the classics section of virtually any video store, and you will find a propaganda piece titled *Inherit the Wind.* It features a fictionalized account of the 1925 Scopes trial, in which creationists are portrayed as bigoted ignoramuses while evolutionists are pictured as benevolent intellectuals. In chapter 5, we will test the theory of evolution in light of reason and empirical science rather than rhetoric and emotional stereotypes. As you will discover, the basic laws of science, including the laws of *effects* and their causes, *energy conservation,* and *entropy* undergird the creation model for origins and undermine the evolutionary model.

Recapitulation

Up to this point we have used FACE as an acronym to equip you to remember the arguments for responding to the farce of evolution. In chapter 6 we will insert the letter *R* and change F-A-C-E into F-A-R-C-E. *R* here equals Recapitulation. This letter will remind you of the recapitulation theory, commonly referred to by the evolutionary cliché, "Ontogeny recapitulates phylogeny." This notion, championed by men such as Carl Sagan, suggests that in the course of an embryo's development (ontogeny), it repeats (recapitulates) the evolutionary history of its species (phylogeny). Thus, at various points the emerging embryo is a fish, then a frog, and finally a fetus. This theory not only relies on *revisionistic history* but has also been used as justification for *Roe v. Wade* and for *racism.*

Epilogue

It is not enough to demonstrate that evolution is a farce. In the epilogue you will be equipped to use your well-reasoned answers as opportunities for sharing the good news that the God who created us also desires reconciliation and fellowship with us. While you cannot change anyone's heart—only the Holy Spirit can do that—you can prepare yourself to sensitively and effectively communicate the difference between mere religion and a meaningful relationship with the Creator.

FACE FARCE

Fossil Follies

- Pseudosaurs
- Pro-Avis
- Punctuated Equilibrium

Ape-Men Fiction, Fraud, & Fantasies

- Pithecanthropus Erectus
- Piltdown Man
- Peking Man

Recapitulation

- Revisionism
- Roe v. Wade
- Racism

Chance

- Eye
- Egg
- Earth

Empirical Science

- Effects
- Energy Conservation
- Entropy

[Evolution] is a general postulate to which all theories, all hypotheses, all systems must henceforward bow and which they must satisfy in order to be thinkable and true. Evolution is a light which illuminates all facts, a trajectory which all lines of thought must follow.

—Pierre Teilhard de Chardin

God said, "Let the land produce vegetation: seed-bearing plants and trees on the land that bear fruit with seed in it, according to their various kinds." And it was so. . . . And God said, "Let the land produce living creatures according to their kinds. . . . And it was so. God made the wild animals according to their kinds, the livestock according to their kinds, and all the creatures that move along the ground according to their kinds.

—the prophet Moses

—Video: Courtesy of Eden Communications

Darwin's study—site of an intellectual revolution.

CHAPTER 1

Truth or Consequences

Other than Scripture, Darwin's magnum opus, *The Origin of Species by Means of Natural Selection,* might well be the most significant literary work in the annals of recorded history. Sir Julian Huxley called the evolutionary dogma it spawned "the most powerful and most comprehensive idea that has ever arisen on earth."[1]

Harvard scientist Ernst Mayr said that the Darwinian revolution of 1859 was "perhaps the most fundamental of all intellectual revolutions in the history of mankind."[2] Likewise, Dr. Michael Denton points out that the far-reaching effects of the Darwinian dogma ignited an intellectual revolution more significant than the Copernican and Newtonian revolutions. He goes on to say,

> The triumph of evolution meant the end of the traditional belief in the world as a purposeful created

> order—the so-called teleological outlook which had been predominant in the western world for two millennia. According to Darwin, all the design, order and complexity of life and the eerie purposefulness of living systems were the result of a simple blind random process—natural selection. Before Darwin, men had believed a providential intelligence had imposed its mysterious design upon nature, but now chance ruled supreme. God's will was replaced by the capriciousness of a roulette wheel. The break with the past was complete.[3]

It would be impossible to overstate the significance of Darwinian evolution. As Denton underscores, the twentieth century cannot be comprehended apart from the intellectual revolution the theory produced:

> The social and political currents which have swept the world in the past eighty years would have been impossible without its intellectual sanction. It is ironic to recall that it was the increasingly secular outlook in the nineteenth century which initially eased the way for the acceptance of evolution, while today it is perhaps the Darwinian view of nature more than any other that is responsible for the agnostic and sceptical outlook of the twentieth century. What was once a deduction from materialism has today become its foundation.[4]

In light of this unprecedented impact of Darwinian dogma, it would be reasonable to expect it to be solidly rooted in truth. In reality, as this book will demonstrate conclusively, evolution is rooted in metaphysical contentions and mythological tales. Denton aptly summed up this sentiment when he termed the Darwinian theory of evolution "the great cosmogenic myth of the twentieth century."[5]

The far-reaching consequences of this cosmogenic myth can be felt in "virtually every field—every discipline of study, every level of education, and every area of practice."[6] The most significant consequence, however, is that it undermines the very foundation of Christianity. If indeed evolution is reflective of the laws of science, then Genesis must be reflective of the flaws of Scripture. And if the foundation of Christianity is flawed, the superstructure is destined to fall. Noted atheist G. Richard Bozarth understood this full well when he penned the following words:

> Christianity is—must be!—totally committed to the special creation as described in *Genesis,* and Christianity must fight with its full might, fair or foul, against the theory of evolution. . . . It becomes clear now that the whole justification of Jesus' life and death is predicated on the existence of Adam and the forbidden fruit he and Eve ate. Without the original sin, who needs to be redeemed? Without Adam's fall into a life of constant sin terminated by death, what purpose is there to Christianity? None. . . . What this all means is that Christianity cannot lose the *Genesis* account of creation like it could lose the doctrine of geocentricism and get along. The battle must be waged, for Christianity is fighting for its very life.[7]

While Bozarth predicted the demise of Christianity without Genesis, he might just as well have predicted the demise of civilization without God. Friedrich Nietzsche, who provided the philosophical framework for Hitler's Germany, understood this better than most. Thus, he predicted that the death of God in the nineteenth century would ensure that the twentieth century would be the bloodiest century in human history.[8]

In the final analysis more consequences for society hinge on the issue of human origins than on any other. Among them are the *sovereignty of self,* the *sexual revolution,* and *survival of the fittest.*

Sovereignty of Self

The supposed death of God ushered in an era in which humans proclaimed themselves sovereigns of the universe. Nowhere was this more in evidence than at the Darwinian Centennial Convention, which celebrated the hundredth anniversary of the publication of Darwin's *The Origin of Species by Means of Natural Selection.* With pomp and ceremony Sir Julian Huxley, the great-grandson of Thomas Huxley, Darwin's bulldog, boasted:

> In the evolutionary system of thought there is no longer need or room for the supernatural. The earth was not created; it evolved. So did all the animals and plants that inhabit it, including our human selves, mind and soul, as well as brain and body. So did religion.

—Photo: AP/WIDE WORLD PHOTOS

Charles Darwin, 1875—a century later his legend lives on.

> Evolutionary man can no longer take refuge from his loneliness by creeping for shelter into the arms of a divinized father figure whom he himself has created.[9]

While the evolutionary system of thought was credited for expunging the need for God, in reality it is merely the repackaging of an age-old deception. In the very first book of the Bible, Satan tells Eve that if she eats the forbidden fruit, "Your eyes will be opened, and you will be like God, knowing good and evil" (Gen. 3:5). What Satan was communicating was that Eve could become the final court of arbitration—she could determine what was right and what was wrong.

Humanity's newfound autonomy sacrificed truth on the altar of subjectivism. Ethics and morals were no longer determined on the basis of objective standards but rather by the size and strength of the latest lobby group. With no enduring reference point, societal norms were reduced to a matter of preference.

One of the most devastating consequences of humanity's repackaging of Satan's age-old deception was the sexual revolution.

Sexual Revolution

I once heard that noted evolutionist Sir Julian Huxley was asked why people had so quickly embraced the theory of evolution. He reputedly retorted: "It is because the concept of a Creator-God interferes with our sexual mores. Thus, we have rationalized God out of existence. To us, He has become nothing more than the faint and disappearing smile of the cosmic Cheshire cat in *Alice in Wonderland*."[10] His response eloquently captures the spirit of the evolutionary paradigm.

With God relegated to the status of a Disney character, we grabbed for all the gusto we could. And what we got in return was adultery, abortion, and AIDS.

Adultery has become commonplace as people fixate on feelings instead of fidelity. As a result, nearly half of all new marriages end in divorce.[11] Abortion has become epidemic as people embrace expediency over ethics. In America alone the death toll for preborn children has exceeded thirty million.[12] AIDS has become pandemic. As people clamor for condoms rather than for commitment, more people have died worldwide from AIDS than America has lost in all of its wars combined.[13]

—Photo: AP/WIDE WORLD PHOTOS

"The concept of a Creator-god interfered with our sexual mores."

Ironically, while prophylactics contribute to the problem, they have been positioned as the key to prevention. In the government-sponsored radio advertisement "Naked," lead singer Anthony Kiedis of the Red Hot Chili Peppers says:

> I've been naked on stage. I've been naked on magazine covers . . . I was born naked and of course, I'm naked whenever I have sex. But now I'm on the radio.
>
> [Disrobing sounds]
>
> So I might as well get naked again. There, I'm naked, see? And what I have here is a condom. A latex condom. I wear one whenever I have sex. Not whenever it's convenient. Or whenever my partner thinks of it. Every time. Look, they're very easy to open.

> [Sounds of package opening]
>
> A breeze to put on. And best of all, they stop the spread of HIV.
>
> Now, I'm naked. With a condom. But I'm not sayin' you should have sex, and I'm not saying you shouldn't have sex. But I'm saying wear a latex condom if you're gonna have sex. Just think of this helpful demonstration and remember: You can be naked without being exposed.[14]

Not only has sex been glorified in the media, it has also been glorified in movies, through music, and by Madison Avenue. Only one rule seems to endure: Life has no rules.

Attempting to rationalize God out of existence in order to do away with His laws of morality is as absurd as voting to repeal the law of gravity because people have fallen off of buildings, bridges, or boats. Even a unanimous vote could not change the deadly consequences for someone who later attempts to jump off of a ten-story building. We cannot violate God's physical or moral laws without suffering disillusionment, destruction, and even death.

Survival of the Fittest

Evolution not only dispenses with God and attempts to make humans the center of the universe, but evolutionism is racist as well. Consider the following excerpt from a letter written by Charles Darwin in 1881:

> The more civilized so-called *Caucasian races* have beaten the Turkish hollow in the struggle for exis-

> tence. Looking to the world at no very distant date, what an endless number of the *lower races* will have been *eliminated* by the *higher civilized races* throughout the world.[15]

Lest this be considered merely an aberration, note that Darwin repeated this sentiment in his book *The Descent of Man.* He speculated, "At some future period, not very distant as measured by centuries, the civilized races of man will almost certainly exterminate, and replace, the savage races throughout the world."[16] In addition, he subtitled his magnum opus *The Preservation of Favored Races in the Struggle for Life.*[17]

And Darwin was not alone in his racist ideology. Thomas Huxley, who coined the term *agnostic*[18] and was the man most responsible for advancing Darwinian doctrine, argued that

> no rational man, cognizant of the facts, believes that the average Negro is the *equal, still less the superior,* of the white man. . . . It is simply incredible [to think] that . . . he will be able to compete successfully with his *bigger-brained and smaller-jawed rival,* in a contest which is to be carried on by *thoughts* and not by *bites.*[19]

Huxley was not only militantly racist but also lectured frequently against the resurrection of Jesus Christ, in whom "[we] are all one" (Gal. 3:28). In sharp distinction to the writings of such noted evolutionists as Hrdlicka, Haeckel, and Hooton,[20] biblical Christianity makes it crystal clear

—Photo: AP/WIDE WORLD PHOTOS

"At some future period, not very distant as measured by centuries, the civilized races of man will almost certainly exterminate and replace the savage races throughout the world."—Charles Darwin

that in Christ "there is neither Jew nor Greek, slave nor free, male nor female" (Gal. 3:28).[21]

In Christianity we sing, "Red and yellow, black and white, all are precious in His sight, Jesus loves the little children of the world."[22] In the evolutionary hierarchy, blacks are placed at the bottom, yellows and reds somewhere in the middle, and whites on top. As H. F. Osborn, director of the American Museum of National History and one of the most prominent American anthropologists of the first half of the twentieth century, put it:

> The Negroid stock is even more ancient than the Caucasian and the Mongolian, as may be proved by

> an examination not only of the brain, of the hair, of the bodily characters, such as the teeth, the genitalia, the sense organs, but of the instincts, the intelligence. The standard of intelligence of the average adult Negro is similar to that of the eleven-year-old youth of the species *Homo sapiens.*[23]

It is significant to note that some of the Crusaders and others who used force to further their creeds in the name of God were acting in direct opposition to the teachings of Christ. The teachings of Osborn and others like him, however, are completely consistent with the teachings of Darwin. Indeed, social Darwinism has provided the scientific substructure for some of the most significant atrocities in human history.

For evolution to succeed, it is as crucial that the unfit die as that the fittest survive. Marvin Lubenow graphically portrays the ghastly consequences of such beliefs in his book *Bones of Contention:*

> If the unfit survived indefinitely, they would continue to "infect" the fit with their less fit genes. The result is that the more fit genes would be diluted and compromised by the less fit genes, and evolution could not take place. The concept of evolution demands death. Death is thus as *natural* to evolution as it is *foreign* to biblical creation. The Bible teaches that death is a "foreigner," a condition superimposed upon humans and nature after creation.[24]

Adolf Hitler's philosophy that Jews were subhuman and that Aryans were supermen led to the extermination of six million Jews. In the words of Sir Arthur Keith, a militant anti-Christian physical anthropologist: "The German Fuhrer, as I have consistently maintained, is an evolutionist; he has consistently sought to make the practices of Germany conform to the theory of evolution."[25]

Karl Marx, the father of communism, saw in Darwinism the scientific and sociological support for an economic experiment that eclipsed even the carnage of Hitler's Germany. His hatred of Christ and Christianity led to the mass murder of multiplied millions worldwide. Karl Marx so revered Darwin that his desire was to dedicate a portion of *Das Kapital* to him.[26]

Sigmund Freud, the founder of modern psychology, was also a faithful follower of Charles Darwin. His belief that man was merely a sophisticated animal led him to postulate that "mental disorders were the vestiges of behavior that had been appropriate in earlier stages of evolution."[27] Daniel Goleman points out that "the evolutionary idea that Freud relied on . . . is the maxim that 'ontogeny recapitulates phylogeny,' that is, that the development of the individual recapitulates the evolution of the entire species."[28] This notion, which I refute in chapter 6, supposes that the conceptus at one stage is a fish rather than a fetus (from the Latin for "infant") and is thus expendable. The human carnage that has resulted from this evolutionary dogma has eclipsed the atrocities of Hitler and Marx combined.

It should also be noted that Darwinian evolution is not only racist but sexist as well. Under the subheading "Difference in the Mental Powers of the Two Sexes," Darwin attempted to persuade followers that

> the chief distinction in the intellectual powers of the two sexes is shewn by man's attaining to a higher eminence, in whatever he takes up, than can woman—whether requiring deep thought, reason, or imagination, or merely the use of the senses and hands. . . . We may also infer . . . [that] the average of mental power in man must be above that of woman.[29]

In sharp contrast to the evolutionary dogma, Scripture makes it clear that all humanity is created in the image of God (Gen. 1:27; Acts 17:29); that there is essential equality between the sexes (Gal. 3:28); and that slavery is as repugnant to God as murder and adultery (1 Tim. 1:10).[30] The consistent application of biblical principles inevitably leads to emancipation. The consistent application of evolutionary principles inevitably leads to enslavement.

The tragic consequences of evolution can hardly be overstated. Denton correctly points out that it "is one of the most spectacular examples in history of how a highly speculative idea for which there is no really hard scientific evidence can come to fashion the thinking of a whole society and dominate the outlook of an age." Furthermore, says Denton,

> Considering its historic significance and the social and moral transformation it caused in western thought, one might have hoped that Darwinian theory was capable of a complete, comprehensive and entirely plausible explanation for all biological phenomena from the origin of life on through all its diverse manifestations up to, and including, the intellect of man.

> That it is neither fully plausible, nor comprehensive, is deeply troubling. One might have expected that a theory of such cardinal importance, a theory that literally changed the world, would have been something more than metaphysics, something more than a myth.[31]

In light of the tragic consequences, it is incredible to think that evolution is still being touted today as truth. The responsibility of demonstrating that it is in reality a farce can no longer be left to a few hired guns in the bastions of higher learning. It is crucial that all thinking human beings become involved in the process as well. This is why I developed the acronym FACE—to make it easy for anyone to remember how to demonstrate the farce of evolution.

We begin with the letter *F*, representing fossil follies. Darwin had predicted that the fossil record would bear him out. In reality, the fossil record has become one of the greatest embarrassments to his legacy.

Why then is not every geological formation and every stratum full of such intermediate links? Geology assuredly does not reveal any such finely-graduated organic chain; and this, perhaps, is the most obvious and serious objection which can be urged against the theory. The explanation lies, as I believe, in the extreme imperfection of the geological record.

—Charles Darwin

We are now about 120 years after Darwin and the knowledge of the fossil record has been greatly expanded. We now have a quarter of a million fossil species, but the situation hasn't changed much. . . . *We have even fewer examples of evolutionary transition than we had in Darwin's time.* (emphasis added)

—David Raup

—Photo: The Natural History Museum, London

The British Museum of Natural History, which houses the world's largest fossil collection—sixty million specimens.

CHAPTER 2

Fossil Follies

Colin Patterson, senior paleontologist at the prestigious British Museum of Natural History, which houses the world's largest fossil collection—sixty million specimens—confessed, "If I knew of any [evolutionary transitions], fossil or living, I would certainly have included them [in my book *Evolution*]."[1] His statement underscores the fact that the fossil record is an embarrassment to evolutionists. No verifiable transitions from one species[2] to another have as yet been found.[3]

Darwin had an excuse. In his day fossil finds were relatively scarce. Today, however, more than a century after his death (one would surmise he is now a creationist), we have an abundance of fossils. Still, we have yet to find even one legitimate transition from one species to another. David Raup, curator of the Field Museum of Natural History in Chicago, underscores this fact:

> We are now about 120 years after Darwin and the knowledge of the fossil record has been greatly expanded. We now have a quarter of a million fossil species, but the situation hasn't changed much. . . . *We have even fewer examples of evolutionary transition than we had in Darwin's time.*[4]

Ironically, while the general public seems blithely unaware that no transitions from one species to another (known as macroevolution[5]) exist, it is common knowledge among paleontologists. That is precisely why novel theories involving *pseudosaurs, pro-avises,* and *punctuated equilibrium* are constantly evolving.

*Pseudosaurs (*Archaeopteryx*)*

Whenever I say that there are no transitions from one species to another, someone inevitably brings up *Archaeopteryx.* This happens so frequently that I've decided to coin a word for the experience: *pseudosaur. Pseudo* means false and *saur* refers to a dinosaur or reptile (literally lizard). Thus, a pseudosaur is a false link between reptiles (such as dinosaurs) and birds. The notion that birds evolved from dinosaurs has so permeated modern thinking that it is "commonly maintained that dinosaurs still survive today in the form of birds, their feathered offspring."[6]

Archaeopteryx—literally "ancient wing"—is said to have twenty-one specialized characteristics in common with particular kinds of dinosaurs.[7] However, as Dr. Duane Gish explains, careful examination has demonstrated that in every case these characteristics are genuinely birdlike

—*Photo: Courtesy of* The Illustrated Origins Answer Book, *Eden Comminications*

Archaeopteryx—the quintessential pseudosaur.

rather than reptilian.[8] Myriad evidences demonstrate conclusively that *Archaeopteryx* is a full-fledged bird, not a missing link. Here are just a few.

First, fossils of both *Archaeopteryx* and the kinds of dinosaurs *Archaeopteryx* supposedly descended from have been found in a fine-grained German limestone formation said to be Late Jurassic (the Jurassic period is said to have begun 190 million years ago, lasting 54 million years).[9] Thus, *Archaeopteryx* is not a likely candidate as the missing link, since birds and their alleged ancestral dinosaurs thrived during the same period of time.[10] In addition, it should be noted that a great deal of controversy has occurred in the evolutionary community as a result of other bird fossils found in sediments classified by evolutionists as Late Triassic (*prior* to the Jurassic). According to this hypothesis, these birds would have lived approximately 75 million years earlier than *Archaeopteryx* and, in fact, at the same time as the first dinosaurs.[11]

Furthermore, initial *Archaeopteryx* fossil finds gave no evidence of a bony sternum, which led paleontologists to conclude that *Archaeopteryx* could not fly or was a poor flyer.[12] However, in April 1993, a seventh specimen was reported that included a bony sternum. Thus, there is no further doubt that *Archaeopteryx* was as suited for power flying as any modern bird.[13] As noted in the highly regarded journal *Science:* "*Archaeopteryx* probably cannot tell us much about the early origins of feathers and flight in true protobirds because *Archaeopteryx* was, in a modern sense, a bird."[14]

Finally, to say that *Archaeopteryx* is a missing link between reptiles and birds, one must believe that scales evolved into feathers for flight. Two basic theories have been set forth by evolutionists to bolster this notion.[15] The arboreal theory holds that wings and feathers developed in tree-dwelling reptiles that jumped from the treetops to escape enemies or to pursue food. Thus, they developed as a mechanism to help ease the animals' fall. The cursorial theory, on the other hand, views ancestral reptiles as ground dwellers that relied on speed for protection or to chase prey. The development of feathered wings and tails lessened wind resistance and supplied lift to increase their speed.

In either case, air friction acting on genetic mutation supposedly frayed the outer edges of reptilian scales. Thus, in the course of millions of years, scales became more and more like feathers until, one day, the perfect feather emerged. To say the least, this idea must stretch the credulity of even the most ardent evolutionists. Consider for a moment the meticulous engineering of feathers. Each is a masterpiece of detail and design.

The central shaft of a feather has a series of barbs projecting from each side at right angles. Rows of smaller barbules in turn protrude from both sides of the barbs. Tiny hooks, called barbicels, project downward from one side of the barbules and interlock with ridges on the opposite side of adjacent barbules. In some feathers there may be as many as a million barbules cooperating to bind the barbs into a complete feather, impervious to air penetration.[16] Furthermore, the positioning of the feathers is controlled by a complex network of tendons that allow them to open like the slats of a blind when the wing is raised. As a result, wind resistance is greatly reduced on the upstroke. On the downstroke, the feathers close, providing resistance for efficient flight. The fearsome flight of the falcon and the delicate, darting flight of the hummingbird clearly illustrate the profound aerodynamic properties of the feathered aerofoil.[17]

These and a myriad of other factors overwhelmingly exclude *Archaeopteryx* as a missing link between birds and reptiles. The sober fact is that *Archaeopteryx* appears abruptly in the fossil record, with masterfully engineered wings and feathers common in the birds observable today. As noted by evolutionist Pierre Lecomte du Nouy, an expert in the science of statistical probability:

> We are not even authorized to consider the exceptional case of the *Archaeopteryx* as a true link. By link, we mean a necessary stage of transition between classes such as reptiles and birds, or between smaller groups. An animal displaying characters belonging to two

> different groups cannot be treated as a true link as long as the intermediary stages have not been found, and as long as the mechanisms of transition remain unknown.[18]

Likewise, Stephen Jay Gould of Harvard University and Niles Eldridge of the American Museum of Natural History, both militant anticreationists, conclude that *Archaeopteryx* cannot be viewed as a transitional form. Here's how they put it:

> At the higher level of evolutionary transition between morphological designs, gradualism has always been in trouble, though it remains the "official" position of most western evolutionists. Smooth intermediates between *bauplane* [basically different types of creatures] are almost impossible to construct, even in thought experiments; there is certainly no evidence for them in the fossil record (curious mosaics like *Archaeopteryx* do not count).[19]

Pro-Avis

A few years after Harvard's Gould ruled out *Archaeopteryx* as a missing link, Yale's John Ostrom proposed a brand-new candidate called pro-avis. Unlike *Archaeopteryx*, no fossil evidence exists for pro-avis.[20] Since science could not produce, science fiction prevailed. Ostrom pictured a prototype in *American Scientist,* and pro-avis sprang into existence.[21]

The pro-avis fairy tale, like any good fairy tale, begins long, long ago with little pro-avises running around on two legs while they playfully caught insects in their scaly little hands. One fateful day an ugly little pro-avis we'll call Mike was born. Unlike his brothers and sisters, little Mikey had frayed scales on both of his hands. Because of little Mikey's imperfections, no one wanted to play with him. Sadly, he had to run around all by himself trying to catch insects. Suddenly, little Mikey discovered something miraculous. Insects stuck like magic to his frayed scales. The more he caught, the better he ate. The better he ate, the faster he ran. The faster he ran, the more his scales frayed. In time little Mikey's ugly scales were transformed into beautiful flying feathers. Soon little Mikey was able to catch insects that would normally have been beyond his reach. It wasn't long before all the little pro-avises wanted to be just like Mike. They began fraying their scales and in time, like Mike, their scales were transformed into fantastic flying feathers. And they lived happily ever after.

In fairy tales, frayed scales turn into feathers, and frogs turn into princes. In evolution all you have to do is add millions of years and little pro-avises turn into beautiful flying birds.[22]

Even the most doctrinaire evolutionists have come to the realization that fairy tales about pseudosaurs like *Archaeopteryx* and pro-avis simply won't fly in an age of scientific enlightenment. *Newsweek* summarized the sentiments of leading evolutionists gathered together at a conference in Chicago as follows: "Evidence from fossils now points overwhelmingly away from the classical Darwinism which most Americans learned in high school."[23] Rather than becoming creationists, however, evolutionists have simply become more creative.[24]

—Illustration: Kristen Ross

Pro-avis—since science could not produce, science fiction prevailed.

Punctuated Equilibrium

The newest fairy tale produced and proliferated in the evolutionary community is a theory called punctuated equilibrium. Its genesis appears to be found in the "hopeful monster" theory of a German geneticist named Richard Goldschmidt.[25] Realizing that there was no compelling evidence for evolution in the fossil record, Goldschmidt speculated that there must have been quantum leaps from one species to another. In his book *The Material Basis of Evolution,* he sums up his sentiments as follows, "The major evolutionary advances must have taken place in single large steps. . . . The many missing links in the paleontological record are sought for in vain because they have never existed: 'the first bird hatched from a reptilian egg.'"[26]

Since I am the father of eight children, Goldschmidt's hopeful monster theory should give me considerable concern. Each time my wife, Kathy, "infanticipates," I should hold down the hospital window lest when she gives birth, our new offspring tries to fly away. Obviously, this is not science; it's science fiction.[27] Tragically, however, this evolutionary fiction is being pawned off on impressionable children as though it were evolutionary fact.

In *The Wonderful Egg,* a book written for children, a mother dinosaur lays an egg that hatches into the very first bird. After growing up into a beautiful specimen replete with wings and feathers, it flies up into a tall tree and sings a happy song. The real tragedy is not that the little bird's song may well become a funeral dirge when it realizes it has no one with which to produce offspring. The real tragedy is that this book earned recommendation from by the prestigious American Association for the Advancement of Science, the American Council on Education, and the Association for Childhood Education International.[28]

Even more tragic is the fact that Gould, one of America's most widely read evolutionists, has come to the defense of Goldschmidt's nonsensical notion. In "The Return of Hopeful Monsters" Gould recounts the "official rebuke and derision" directed at Goldschmidt by the scientific establishment. He then predicts "that during the next decade Goldschmidt will be largely vindicated in the world

—Illustration: Kristen Ross

The first bird hatched from a reptilian egg.

of evolutionary biology."[29] Although his prediction has not come to pass, it has not stopped him from propagating a theory called punctuated equilibrium, which in reality is little more than a regurgitation of Goldschmidt's hopeful monster theory.[30]

According to Gould's theory of punctuated equilibrium, "a species does not arise gradually by the steady transformation of its ancestors; it appears all at once and 'fully formed.'"[31] As Gish explains, punctuated equilibrium proponents speculate that

> once a species has developed, it proliferates into a large population and persists relatively unchanged for one, two, five, or ten million years, or even longer. Then for some unknown reason a relatively small number of the individuals of the population become isolated, and by some unknown mechanism rapidly evolve into a new species (by rapid is meant something on the order of tens of thousands of years). Once the new species has evolved, it then either becomes rapidly extinct or proliferates into a large population. This large population then persists for one or more millions of years. The long period of stasis is the portion of the process referred to as the period of equilibrium, and the interval characterized by rapid evolution is the punctuation—thus the term, punctuated equilibrium.[32]

The problems with punctuated equilibrium should be self-evident. First, this theory is motivated by the lack of vertical transitional forms in the fossil record. Thus, it is a classic argument from silence. Furthermore, this convoluted concept flies in the face of the science of genetics. As noted by Gish,

> The genetic apparatus of a lizard, for example, is devoted 100% to producing another lizard. The idea that this indescribably complex, finely tuned, highly integrated, amazingly stable genetic apparatus

> involving hundreds of thousands of interdependent genes could be drastically altered and rapidly reintegrated in such a way that the new organism not only survives but actually is an improvement over the preceding form is contrary to what we know about the apparatus and how it functions.[33]

Finally, as noted by Drs. Henry Morris and Gary Parker, the effects produced by "jumping genes" or chromosomal rearrangements would not produce a hopeful monster but a monstrosity. Even if, as postulated by punctuated equilibrium, the jumps do not go from reptiles to birds but from scales to feathers, the jumps are still too fantastic. Conversely, if the jumps are in reality rather insignificant, then we are right back at square one—gradualism.[34] And as noted by Gould, "The extreme rarity of transitional forms persists as the trade secret of paleontology."[35]

Creationists and Christians should be grateful for the candor of doctrinaire evolutionist Colin Patterson when he admitted:

> For over 20 years I thought I was working on evolution. . . . [But] there was not one thing I knew about it. . . . So for the last few weeks I've tried putting a simple question to various people and groups of people. Question is: "Can you tell me anything you know about evolution, any one thing, any one thing that is true?" I tried that question on the geology staff at the Field Museum of Natural History and the only answer I got was silence. I tried it on the members of the Evolutionary Morphology

> Seminar in the University of Chicago, a very prestigious body of evolutionists, and all I got there was silence for a long time and eventually one person said, "Yes, I do know one thing—it ought not to be taught in high school." . . . During the past few years . . . you have experienced a shift from evolution as knowledge to evolution as faith. . . . Evolution not only conveys no knowledge, but seems somehow to convey anti-knowledge.[36]

Nowhere is this antiknowledge more readily apparent than in the fiction, fraud, and fantasy surrounding ape-men. In chapter 3 we will use the *A* in the acronym FACE to remind you that humans are made in the image of the Almighty, not the image of apes.

If you go back far enough, we and the chimps share a common ancestor. My father's father's father's father, going back maybe a half million generations—about five million years—was an ape.

—*Ape Man: The Story of Human Evolution,*
hosted by Walter Cronkite

God created man in his own image, in the image of God he created him; male and female he created them.

—the prophet Moses

—Photo: *The Illustrated London News Picture Library*

A single tooth is transformed by an artist into Nebraska Man and Nebraska Mom—as appeared in *The Illustrated London News* June 24, 1922.

CHAPTER 3

Ape-Men Fiction, Fraud, and Fantasy

In 1922, a tooth was discovered in Nebraska. With a little imagination the tooth was connected to a mythological jawbone, the jawbone was connected to a skull, the skull was connected to a skeleton, and the skeleton was given a face, features, and fur. By the time the story hit a London newspaper, not only was there a picture of "Nebraska man" but there was also a picture of "Nebraska mom."[1] All of that from a single solitary tooth. Imagine what might have happened if a skeleton had been discovered. Perhaps a yearbook would have been published!

Some time after the initial discovery, an identical tooth was found by geologist Harold Cook. This time the tooth was attached to an actual skull, and the skull was attached to the skeleton of a wild pig. Thus, Nebraska man, known by the "scientific" designation *Hesperopithecus haroldcookii,* has been unmasked as a myth rather than a man in the making.

Ironically, while scientists were attempting to make a monkey out of a pig, the pig made a monkey out of the scientists.[2] While one would think this blunder would preclude the possibility of similar fantasies, a parade of pretenders continues to persist.

Pithecanthropus Erectus

Speculation about *Pithecanthropus erectus,* the ape-man that walked erect, is far and away the most famous "ape-man" fiction still being circulated as fact. While over time he has evolved into a new classification called *Homo erectus,* millions regard him as a friendly ancestor, not just a fossil, and simply refer to him by the nickname Java man.

It is generally known that Java man was initially discovered by a Dutchman named Eugene Dubois on the Dutch East Indian island of Java in 1891. What is not so well known is that Java man consists of nothing more than a skullcap, a femur (thigh bone), three teeth, and a great deal of imagination. Even more disturbing is the fact that the femur was found fifty feet from the skullcap and a full year later. Most unsettling of all is that for almost thirty years, Dubois downplayed his discovery of two human skulls (the Wadjak skulls), which he found in close proximity to his original "finds."[3] This alone should have been sufficient to disqualify Java man as humankind's ancestor.

The famous evolutionist, Sir Arthur Keith, drove this point home in humorous fashion:

> If, on his return in 1894, [Dubois] had placed before the anthropologists of the time the ape-like skull from

—Illustration: Kristen Ross

Pithecanthropus erectus: A skullcap, a femur, three teeth, and a great deal of imagination.

Trinil [the skull of Java man] side by side with the great-brained skulls of Wadjak, both fossilized, both from the same region of Java, he would have given them a meal beyond the powers of their mental digestion. Since then our digestions have grown stronger.[4]

Keith, of course, was speaking tongue-in-cheek to underscore the fact that those in his own profession have become increasingly gullible. In truth, the most thorough fact-finding expedition ever conducted on Java man utterly demolished Dubois's claims. This trek, commonly referred to as the Selenka Expedition, included nineteen evolutionists bent on demonstrating that the evolutionary conjectures about Java man were true. However, their 342-page scientific report, which, according to Keith, "commands our unstinted praise," demonstrates beyond the peradventure of a doubt that Java man played no part in human evolution.[5]

Despite all the evidence, it is truly amazing that *Time* magazine printed "How Man Began,"[6] an article that shamelessly treated Java man as though it were a true evolutionary ancestor. Even more incredible is the fact that Donald Johanson, best known for his discovery of a famous fossil named Lucy (after the Beatles' tune, "Lucy in the Sky with Diamonds"), still regards Java man as a valid transitional form; and Harvard's Richard Lewontin thinks this information about Java man should be taught as one of five "facts of evolution."[7]

Piltdown Man

While *Pithecanthropus erectus* (Java man) can best be placed in the category of fiction, Piltdown man *(Eoanthropus dawsoni)* may be factually described as a fraud. While the fraud may have been cleverly conceived, it was crudely carried out. The jaw of an ape was stained to make it appear as though it matched a human skull; the Piltdown fossils along with accompanying bones were not

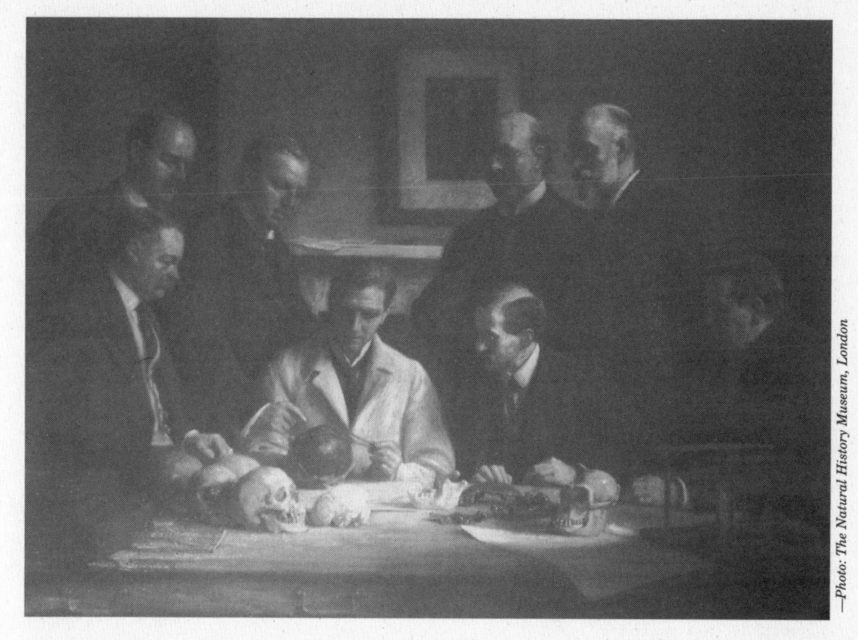

—Photo: The Natural History Museum, London

The "Piltdown Gang" by John Cook. Sir Arthur Keith (center, wearing a lab coat). Charles Dawson (second from right), A. S. Woodward (far right).

only stained but also reshaped.[8] As Marvin Lubenow explains:

> The file marks on the orangutan teeth of the lower jaw were clearly visible. The molars were misaligned and filed at two different angles. The canine tooth had been filed at two different angles. The canine tooth had been filed down so far that the pulp cavity had been exposed and then plugged.[9]

Despite the fact that the Piltdown fossils were clearly "doctored," highly esteemed scientists in the field affirmed their veracity. William Fix notes in *The Bone Peddlers* that the two most eminent paleoanthropologists in England at the time, Sir Arthur Keith and A. S. Woodward, declared that Piltdown man "represents more closely than any human form yet discovered the common ancestor from which both the Neanderthal and modern types have been derived."[10]

It wasn't until 1953, after the Nature Conservancy had spent a considerable amount of taxpayer money to designate the Piltdown site as a national monument, that Dawson's Dawn man (Piltdown) was formally declared a fake.[11] Although there is still a great deal of uncertainty as to who perpetrated the fraud, A. S. Woodward (keeper of geology at the British Museum), Pierre Teilhard de Chardin (a Jesuit priest), and Charles Dawson (the lawyer who unearthed Piltdown man in 1912) are front-runners in the long list of possible suspects.[12]

While Piltdown man may well be ranked as one of the more notorious scientific frauds in history, it was used for forty years to dupe unsuspecting students into thinking that evolution was a fact.

Peking Man[13]

While Java man is fictitious and Piltdown man is a fraud, Peking man might best be described as pure fantasy. Like Nebraska man, Peking man was based originally on a dusty old tooth. It was conveniently discovered in China, just as Canadian physician Davidson Black was about to run out of funds for his evolutionary explorations in 1927.[14]

The Rockefeller Foundation rewarded this discovery with a generous grant, permitting Black to continue digging. Two years later, he discovered what he fervently believed was Peking man's braincase, and he estimated Peking man to be half a million years old. Unfortunately, Black's fame was fleeting, for at age forty-nine, he died of a heart attack.[15]

Black's death, however, did not end his dreams. By the time World War II broke out, the evolutionary community had "discovered" fourteen skulls and an interesting collection of tools and teeth. All fourteen skulls were "missing in action" by war's end, yet the pretense persisted.[16]

The photographs and plaster casts that remained had some interesting similarities. Apart from the fact that the lower skeletons were missing, the skulls had all been bashed at the base. As Ian Taylor points out, Teilhard de Chardin of Piltdown fame made his former professor, Marcellin Boule, angry "at having traveled halfway around the world to see a battered monkey skull. He pointed out that all the evidence indicated that true man was in charge of some sort of 'industry' and that the skulls found were merely those of monkeys."[17]

Boule was not far from the truth. As Gish has pointed out in debates against evolutionists, it now seems likely that the tools found with Peking man were used *on* him, not *by* him.[18] As it turns out, while monkey meat is difficult to digest, monkey brains are delicious. To this day, natives of Southeast Asia lop off the heads of monkeys, bash them in at the back, scoop out the brains, and eat them as a delicacy. If you saw the movie *Indiana Jones: Temple of Doom,* that's exactly what Jones and his cohorts had for dinner—"Peking man on the half shell." It is now clear to anyone who looks at the evidence with an open mind that Peking man was not a distant relative but rather dinner.

To say that "hominids" like Peking man and his partners

—Photo: The Natural History Museum, London

Peking Man: Another product of a fertile imagination.

are closely related to humans because both can walk is like saying that a hummingbird and a helicopter are closely related because both can fly. In reality, the distance between an ape who cannot read or write and a descendant of Adam who can compose a musical masterpiece or send a person to the moon is the distance of infinity.

All the evidence in the world, however, is not sufficient to convince those who do not want to be confused with facts.[19] To wit, Walter Cronkite, in the television premiere of

Ape Man: The Story of Human Evolution, declared that monkeys were his "newfound cousins." Cronkite went on to say: "If you go back far enough, we and the chimps share a common ancestor. My father's father's father's father, going back maybe a half million generations—about five million years—was an ape."[20]

In fairness it should be pointed out that not all evolutionists believe humans evolved from monkeys. Some in fact believe quite the opposite—*that monkeys evolved from humans.* Geoffrey Bourne, former director of the Yerkes Primate Research Center of Emory University in Atlanta, Georgia, is a classic case in point. An article in *Modern People* points out that Bourne, who "is considered one of the world's leading experts on primates," believes that "monkeys, apes, and all other lower primate species are really the offspring of man."[21]

Bourne's beliefs are bolstered by an article in *New Scientist* in which John Gribbin and Jeremy Cherfas say they "think that the chimp is descended from man." Their theory is that "the genetic changes that produced early man from an ape were cleanly reversed to produce early chimps and gorillas from man."[22] The truth, however, is that evolutionists who believe humans evolved from chimps over millions of years, as well as those who believe chimps evolved from humans, are dead wrong.

No matter how many years the evolutionist postulates, chance operating on natural processes can no more create a chimp than it could create a cell. With that in mind, let's proceed to the letter *C* in the acronym FACE, which demonstrates that chance renders evolution not only improbable but indeed impossible.

Chance *alone* is at the source of every innovation, of all creation in the biosphere. Pure chance, absolutely free but blind, at the very root of the stupendous edifice of evolution: this central concept of modern biology is no longer one among other possible or even conceivable hypotheses. It is today the *sole* conceivable hypothesis. . . . And nothing warrants the supposition—or the hope—that on this score our position is likely ever to be revised

—Jacques Monod

Chance, when strictly examined, is a mere negative word, and means not any real power, which has anywhere, a being in nature.

—David Hume

—*Photo: Scala/Art Resource, NY*

Imagine asserting that the *Last Supper* painted itself apart from Leonardo da Vinci (1452-1519).

CHAPTER 4

Chance

One of the primary dilemmas of evolutionary theory is that it forces scientists to conclude that the cosmos in all of its complexity was created by chance. As biologist Jacques Monod, a winner of the prestigious Nobel prize, puts it, "chance *alone* is at the source of every innovation, of all creation in the biosphere. Pure chance, absolutely free but blind, [is] at the very root of the stupendous edifice of evolution."[1] As noted theologian R. C. Sproul explains, for the materialist, chance is the "magic wand to make not only rabbits but entire universes appear out of nothing."[2] Sproul also warns that "if chance exists in any size, shape, or form, God cannot exist. The two are mutually exclusive. If chance existed, it would destroy God's sovereignty. If God is not sovereign, he is not God. If he is not God, he simply *is* not. If chance is, God is not. If God is, chance is not."[3]

Chance in this sense refers to that which happens without cause.[4] Thus, chance implies the absence of both a design and a designer. Reflect for a moment on the absurdity of such a notion. Imagine suggesting that Christopher Wren had nothing whatsoever to do with the design of St. Paul's Cathedral in London. Imagine asserting that the majestic *Messiah* composed itself apart from Handel. Or imagine that the *Last Supper* painted itself without Leonardo da Vinci.

Now consider an even more egregious assertion. Consider the absurdity of boldly asserting that an eye, an egg, or the earth, each in its vast complexity, is merely a function of random chance.[5] Ironically, Darwin himself found it hard to swallow the notion that the eye could be the product of blind evolutionary chance, conceding that the intricacies of the human eye gave him "cold shudders."[6]

Eye

In his landmark publication, *The Origin of Species by Means of Natural Selection,* Darwin avowed,

> To suppose that the eye, with all its inimitable contrivances for adjusting the focus to different distances, for admitting different amounts of light, and for the correction of spherical and chromatic aberration, could have been formed by natural selection, seems, I freely confess, absurd in the highest degree possible.[7]

He labeled this dilemma as the problem of "organs of extreme perfection and complication."[8]

Consider for a moment the incredible complexity of the human eye. It consists of a ball with a lens on one side and a light-sensitive retina made up of rods and cones inside the other. The lens itself has a sturdy protective covering called a cornea and sits over an iris designed to protect the eye from excessive light. The eye contains a fantastic watery substance that is replaced every four hours, while tear glands continuously flush the outside clean. In addition, an eyelid sweeps secretions over the cornea to keep it moist, and eyelashes protect it from dust.[9]

It is one thing to stretch credulity by suggesting that the complexities of the eye evolved by chance; it is quite another to surmise that the eye could have evolved in concert with myriad other coordinated functions. As a case in point, extraordinarily tuned muscles surround the eye for precision motility and shape the lens for the function of focus.[10]

Additionally, consider the fact that as you read this document, a vast number of impulses are traveling from your eyes through millions of nerve fibers that transmit information to a complex computing center in the brain called the visual cortex. Linking the visual information from the eyes to motor centers in the brain is crucial in coordinating a vast number of bodily functions that are axiomatic to the very process of daily living. Without the coordinated development of the eye and the brain in a synergistic fashion, the isolated developments themselves become meaningless and counterproductive.[11]

In *Darwin's Black Box,* biochemist Michael Behe points out that what happens when a photon of light hits a human eye was beyond nineteenth-century science. Thus, to Darwin vision was an unopened black box.[12] In the twentieth century, however, the black box of vision has been opened, and it is no longer enough to consider the anatomical structure of

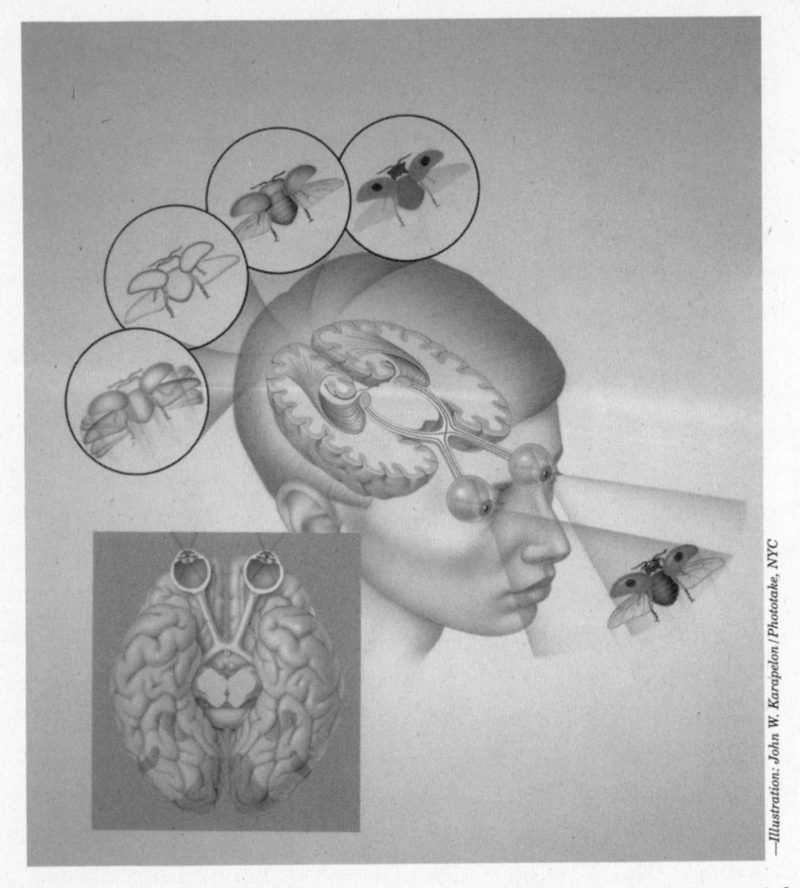

Linking information from the eyes to motor centers in the brain is crucial in coordinating bodily functions axiomatic to the process of daily living.

the eye. We now know that "each of the anatomical steps and structures that Darwin thought were so simple actually involves staggeringly complicated biochemical processes" that demand explanation.[13]

Behe goes on to demonstrate that one cannot explain the origin of vision without first accounting for the origin of the enormously complex system of molecular mechanisms that make it work. The complexity of the biochemistry of vision is such that I've relegated further detail to a note.[14] Phillip Johnson, author of *Defeating Darwinism by Opening Minds,* has aptly summarized Darwin's dilemma regarding the eye: "Evolutionary biologists have been able to pretend to know how complex biological systems originated only because they treated them as black boxes. Now that biochemists have opened the black boxes and seen what is inside, they know the Darwinian theory is just a story, not a scientific explanation."[15]

Egg

In *Darwin's Black Box,* Behe further notes that there are black boxes within black boxes. As science advances, more and more of these black boxes are being opened, revealing an "unanticipated Lilliputian world" of enormous complexity that has pushed the theory of evolution beyond the breaking point.[16] Evolution cannot account for the astonishingly complex synchronization process needed for, say, the shell of an emerging egg to form from the calcium that is stored inside the bones of a bird's body.[17] This shell not only provides a protective covering for the egg but also provides a source of calcium for the emerging embryo and a membrane through which it can breath.[18] Furthermore, evolution cannot account for the complex synchronization process needed to produce life from a single fertilized human egg.

"The tapestry of life begins with a single thread."[19] Through a process of incredible precision, a microscopic egg in one human being is fertilized by a sperm cell from another.

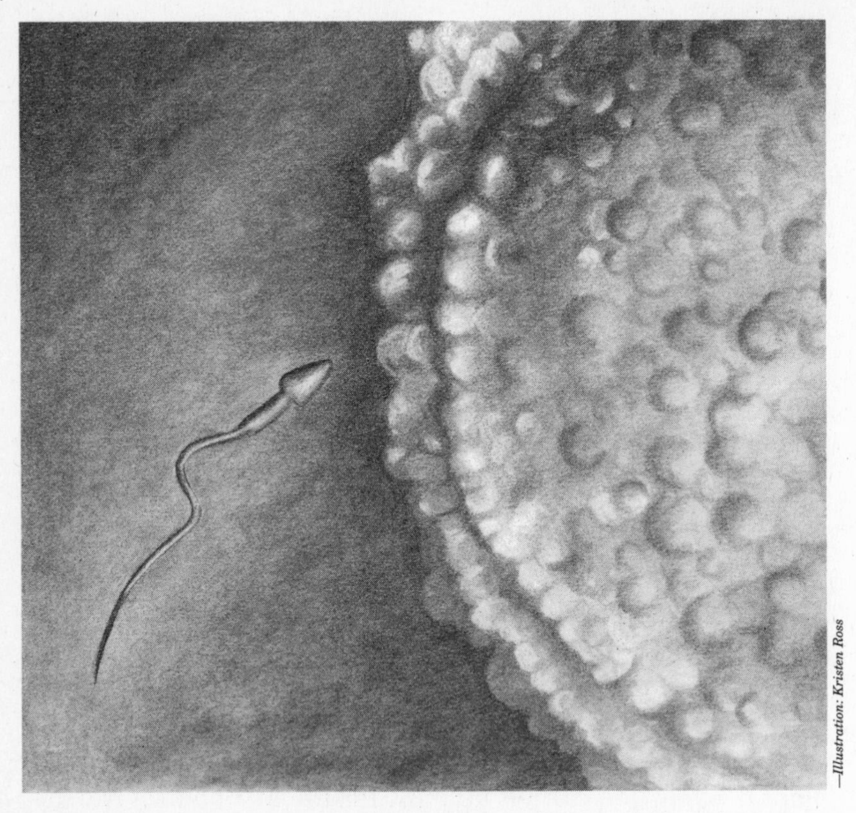

—*Illustration: Kristen Ross*

Through a process of incredible precision a microscopic egg in one human is fertilized by a sperm cell from another.

This process not only marks the beginning of a new life but also marks the genetic future of that life.[20] A single fertilized egg (zygote), the size of a pinhead, contains chemical instructions that would fill more than 500,000 printed pages.[21] The genetic information contained in this "encyclopedia" determines the potential physical aspect of the developing human from height to hair color. In time the fertilized egg divides into the 30 trillion cells that make up the human body, including

12 billion brain cells, which form more than 120 trillion connections.[22]

In Darwin's day, a human egg was thought to be quite simple—for all practical purposes, little more than a microscopic blob of gelatin. Today, we know that a fertilized egg is among the most organized, complex structures in the universe. In an age of scientific enlightenment, it is incredible to think that people are willing to maintain that something so vastly complex arose by chance. As Dr. James Coppedge, an expert in the science of statistical probability, puts it, "Chance requires ten billion tries on the average in order to count to ten."[23]

In an experiment using ten similar coins numbered one through ten, chance will succeed on the average only once in ten billion attempts to get the number one followed in order by all the rest. Coppedge explains that if a person could draw and record one coin every five seconds day and night, it would still take more than 1,500 years for chance, on average, to succeed just once in counting to ten.[24] He goes on to demonstrate the difference intelligence makes by documenting that a *child* can do in minutes what *chance* would take a millennium to do. "Chance really 'doesn't have a chance' when compared with the intelligent purpose of even a child."[25] Even more revealing is the fact that a child playing the party game Scrabble can easily spell the phrase, "the theory of evolution," while chance requires five million times the assumed age of the earth to accomplish the same feat.[26]

Earth

Like an egg or an eye, the earth is a masterpiece of precision and design that could not have come into existence by chance. Astronaut Guy Gardner, who has seen the earth from the perspective of the moon, points out that "the more we learn and see about our universe the more we come to realize that the most ideally suited place for life within the entire solar system is the planet we call home."[27] King David said it best:

> The heavens declare the glory of God;
> the skies proclaim the work of his hands.
> Day after day they pour forth speech;
> night after night they display knowledge.
> There is no speech or language
> where their voice is not heard.
> Their voice goes out into all the earth,
> their words to the ends of the world. (Ps. 19:1–4)

Let's take a few minutes to explore the miracles that demonstrate life on earth was designed by a benevolent Creator rather than directed by blind chance.

First, consider plain old tap water. The solid state of most substances is more dense than their liquid state, but the opposite is true for water, which explains why ice floats rather than sinks. If water were like virtually any other liquid, it would freeze from the bottom up rather than from the top down, killing aquatic life, destroying the oxygen supply, and making earth uninhabitable.[28]

Furthermore, ocean tides, which are caused by the gravitational pull of the moon, play a crucial role in our survival. If the moon were significantly larger, thereby having a stronger gravitational pull, devastating tidal waves would submerge large areas of land. If the moon were smaller,

This photo, released by NASA December 29, 1968, shows the spectacular "earthrise" which greeted Apollo 8 astronauts.

tidal motion would cease, and the oceans would stagnate and die.[29]

Finally, consider the ideal temperatures on planet Earth—not duplicated on any other known planet in the universe. If we were closer to the sun, we would fry. If we were farther away, we would freeze.[30]

From the tap water to the tides and temperatures that we so easily take for granted, the earth is an unparalleled planetary masterpiece. Like Handel's *Messiah* or da Vinci's *Last Supper,* it should never be carelessly pawned off as the result of blind evolutionary processes. Yet, tragically, in an age of scientific enlightenment many are doing just that. Consider the following introduction to *The Miracle of Life,* an Emmy award–winning PBS NOVA broadcast on evolution:

> Four and a half billion years ago, the young planet Earth was a mass of cosmic dust and particles. It was almost completely engulfed by the shallow primordial seas. Powerful winds gathered *random molecules* from the atmosphere. Some were deposited in the seas. Tides and currents swept the *molecules* together. And somewhere in this ancient ocean the miracle of life began. . . . *The first organized form of primitive life was a tiny protozoan [a one-celled animal].* Millions of protozoa populated the ancient seas. These early organisms were completely self-sufficient in their sea-water world. *They moved about their aquatic environment feeding on bacteria and other organisms.* . . . From these one-celled organisms evolved all life on earth.[31]

The *real miracle of life* is how someone could stand for such nonsense in the twentieth century. First, how could the protozoa be the first form of primitive life if there were already organisms such as bacteria in existence? Molecular biology has demonstrated empirically that bacteria are incredibly complex. In the words of Michael Denton,

> Although the tiniest bacterial cells are incredibly small, weighing less than 10^{-12}gms, each is in effect a veritable micro-miniaturized factory containing thousands of exquisitely designed pieces of intricate molecular machinery, made up altogether of one hundred thousand million atoms, far more complicated than any machine built by man and absolutely without parallel in the non-living world.[32]

Furthermore, far from being primitive, the protozoa that were thought to be simple in Darwin's day have been shown by science to be enormously complex. Molecular biology has demonstrated that there is no such thing as a "primitive" cell. To quote Denton again, "In terms of their basic biochemical design . . . no living system can be thought of as being primitive or ancestral with respect to any other system, nor is there the slightest empirical hint of an evolutionary sequence among all the incredibly diverse cells on earth."[33]

Finally, as Coppedge documents, giving evolutionists every possible concession, postulating a primordial sea with every single component necessary, and speeding up the rate of bonding a trillion times:

> The probability of a single protein[34] molecule being arranged by chance is 1 in 10^{161}, using all atoms on earth and allowing all the time since the world began. . . . For a minimum set of the required 239 protein molecules for the smallest theoretical life, the probability is 1 in $10^{119,879}$. It would take $10^{119,841}$ years on the average to get a set of such proteins.

> That is $10^{119,831}$ times the assumed age of the earth and is a figure with 119, 831 zeroes.[35]

To provide a perspective on how enormous a one followed by 161 zeros is, Coppedge uses the illustration of an amoeba (a microscopic one-celled animal) moving the entire universe (including every person, the earth, the solar system, the Milky Way galaxy, millions of other galaxies, etc.) over the width of one universe, atom by atom. This amoeba is going to move the entire universe over one universe (the universe is thirty billion light-years in diameter—to calculate the number of miles multiply thirty billion by 5.9 trillion) at the slowest possible speed. The amoeba is going to move one angstrom unit (the width of a hydrogen atom—the smallest known atom) every fifteen billion years (the supposed age of the universe). Obviously the amoeba would have to move zillions of times before the naked eye could detect that it had moved at all.

At this rate the amoeba travels thirty billion light-years and puts an atom down one universe over. It then travels back at the same rate of speed and takes another atom from your body and moves it one universe over. Once it has moved you over, it moves the next person and then the next until it has moved all five billion or so people on planet Earth. It then moves over all the houses and cars, the solar system, the Milky Way galaxy, and millions of other galaxies that exist in the known universe.

In the time that it took to do all that, we would not get remotely close to forming one protein molecule by random chance.[36] If, however, a protein molecule is eventually formed by chance, forming the second one would be infinitely more difficult. As you can see, the science of statistical probability demonstrates conclusively that forming a protein molecule by random processes is not only improbable; it is impossible. And forming a living cell is beyond illustration. As King David poignantly put it, "The fool says in his heart, 'There is no God'" (Ps. 14:1).

We now move to the letter *E* in the acronym FACE, which represents empirical science. As you proceed you will be equipped to demonstrate that the basic laws of science undermine the theory of evolution and undergird the fact of creation.

If all the achievements of theologians were wiped out tomorrow, would anyone notice the smallest difference? Even the bad achievements of scientists . . . work! The achievements of theologians don't do anything, don't affect anything, don't mean anything. What makes anyone think that "theology" is a subject at all?

—Richard Dawkins

I find it as difficult to understand a scientist who does not acknowledge the presence of a superior rationality behind the existence of the universe as it is to comprehend a theologian who would deny the advances of science. And there is certainly no scientific reason why God cannot retain the same relevance in our modern world that he held before we began probing His creation with telescope, cyclotron, and space vehicles. Our survival here and hereafter depends on adherence to ethical and spiritual values.

—Wernher von Braun

—Photo: AP/WIDE WORLD PHOTOS

The Scopes Trial of 1925 (Creationist lawyer William Jennings Bryan is 6th from the left with left hand to his mouth.

CHAPTER 5

Empirical Science

Go to the classics section of virtually any video store and you will find *Inherit the Wind,* a propaganda piece starring Spencer Tracy and Gene Kelly. It features a fictionalized account of the 1925 Scopes trial, in which a teacher is jailed for violating a state law prohibiting the teaching of evolution. Creationists are portrayed as bigoted ignoramuses while evolutionists are pictured as benevolent intellectuals.[1] In the end one is left with the notion that believing in the creation model for origins is tantamount to committing intellectual suicide. In reality, nothing could be farther from the truth. Some of the greatest intellects the world has ever known were defenders of a creation view of origins.

Leonardo da Vinci, considered by some to be the real founder of modern science, was a committed creationist. Robert Boyle, the father of modern chemistry as well as the greatest physical scientist of his generation, was a great apologist for the Genesis account of origins. Isaac Newton, a prodigious intellect who developed calculus, discovered the law of gravity, and designed the first reflecting telescope, not only refuted atheism but also strongly defended the biblical account of creation. Louis Pasteur, well known for the process of pasteurization and for utterly demolishing the concept of spontaneous generation, was devoutly religious and strongly opposed Darwinian evolution.[2]

Dr. Henry Morris devoted a book to "men of science and men of God," which includes other intellects including Johannes Kepler (scientific astronomy), Francis Bacon (scientific method), Blaise Pascal (mathematician), Carolus Linnaeus (biological taxonomy), Gregor Mendel (genetics), Michael Faraday (electromagnetics), and Joseph Lister (antiseptic surgery).[3] Albert Einstein, one of the greatest intellects of modern times, also "came to the conclusion that God did not create by chance, but rather that he worked according to planned, mathematical, teleonomic, and therefore—to him—rational guidelines."[4]

Rather than falling for rhetoric and emotional stereotypes—such as those presented in *Inherit the Wind*—these men were deeply committed to reason and empirical science. Like them, we would do well to test the theory of evolution in light of the laws of science instead of trying them in the court of public opinion.

Effects and Their Causes

In *2001: A Space Odyssey,* astronauts exploring the moon discovered an obelisk. Moviegoers immediately understood the import of the discovery. They may not have known

—Illustration: Kristen Ross

It doesn't take a rocket scientist to understand that an effect (obelisk) must have a cause (designer) greater than itself.

whether the obelisk was designed by space aliens, angels, or the lost civilization of Atlantis, but they did know that an intelligent being had previously been to the moon. Had the plot of the movie attempted to suggest that the obelisk was an effect without an intelligent cause, it would have been laughed out of theaters.[5]

It doesn't take a rocket scientist to understand that an effect (an obelisk) must have a cause (designer) greater than itself. As Fred Heeren points out, "This is common sense, and no one has ever observed an exception. Even Julie Andrews sings about it: 'Nothing comes from nothing; nothing ever could.' That every effect must have a cause is a self-evident truth, not only for those who have been trained in logic, but for thinking people everywhere."[6]

Cause and effect, "which is universally accepted and followed in every field of science, relates every phenomenon as an effect to a cause. No effect is ever quantitatively 'greater' nor qualitatively 'superior' to its cause. An effect can be lower than its cause but never higher."[7] In stark contrast, the theory of evolution attempts to make effects such as organized complexity, life, and personality greater than their causes—disorder, nonlife, and impersonal forces. As has been well said, "'Teleology is a lady without whom no biologist can exist; yet he is ashamed to be seen with her in public.' Design requires a designer, and this is precisely what is lacking in non-theistic [materialistic] evolution."[8]

In the television series, *Cosmos,* Carl Sagan boldly pontificated—but never proved—his premise that "the cosmos is all that is or ever was or ever will be."[9] In stark contrast, Albert Einstein humbly acknowledged that "the harmony of natural law . . . reveals an intelligence of such superiority that, compared with it, all the systematic thinking and acting of human beings is an utterly insignificant reflection."[10] Long before Albert Einstein, another prodigious intellect, the apostle Paul, said essentially the same thing: "For since the creation of the world God's invisible qualities—his eternal power and divine nature—have been clearly seen, being understood from what has been made, so that men are without excuse" (Rom. 1:20). In commenting on this passage, Fred Heeren says,

> This self-evident truth is the simple, rational deduction that all we see is an *effect* which demands a very great, supernatural Cause. The sun and the stars, the moon and this Earth could not have come from nothing. That's irrational—not just to the Western mind, but to the *human* mind. Every phenomenon in the universe can be explained in terms of something else that caused it. But when the phenomenon in question is the existence of the universe itself, there is nothing in the universe to explain it. No *natural* explanation.[11]

One more point needs to be made before we move on. In saying that the universe is an effect that requires a cause equal to or greater than itself, one may well presume that this principle would apply equally to God. This, however, is clearly not the case. Unlike the universe, which according to modern science had a beginning, God is eternal. Thus, as an eternal being, God can be demonstrated logically to be the uncaused first cause.[12]

Energy Conservation

Today's physical sciences are built on three laws of thermodynamics that describe energy relationships of matter in the universe.[13] These laws were established as a scientific discipline by Lord William Kelvin (1824–1907), a committed Christian, who, like Einstein, was a brilliant intellect. *The New Encyclopedia Britannica* states that he published hundreds of

—Photo: AP/WIDE WORLD PHOTOS

Evolutionist Dr. Isaac Asimov describes Energy Conservation as "the most powerful and most fundamental generalization about the universe that scientists have ever been able to make."

papers, was granted multiple patents, and that it was said of Lord Kelvin that he was "entitled to more letters after his name than any other man in Britain." One of his greatest contributions to science was his role in the development of the law of energy conservation.[14]

Like the law of cause and effect, the law of energy conservation is an empirical law of science. Also known as the first law of thermodynamics, the law of energy conservation states that while energy can be converted from one form to another, it can neither be created nor annihilated.[15] According to Isaac Asimov, it "is considered the most powerful and most fundamental generalization about the universe that scientists have ever been able to make."[16] Fred Heeren notes that the first law of thermodynamics means that

> neither mass nor energy can appear from nothing. Such an occurrence would be "a free lunch," as big bang theorist Alan Guth likes to say, in contradiction to the common sense notion that there is no free lunch. And yet, there is no denying that the universe is here; so the universe itself appears to be a free lunch. But from the laws of physics we see operating today, creation is impossible as an ongoing event. That is, the conditions that we know hold true in our present universe prevent any possibility of matter springing out of nothing today.[17]

From a purely logical point of view it should be self-evident that nothing comes from nothing. In other words, it is illogical to believe that *something* could come from *nothing*. Yet, this is precisely what philosophical naturalism—the worldview undergirding evolutionism—presupposes. This is analogous to the nineteenth-century concept of spontaneous generation, or what Michael Behe humorously refers to as "Calvinism":

> Calvin is always jumping in a box with his stuffed tiger, Hobbes, and traveling back in time, or "transmogrifying" himself into animal shapes, or using it as a "duplicator" and making clones of himself. A little

boy like Calvin easily imagines that a box can fly like an airplane (or something), because Calvin doesn't know how airplanes work.

In some ways, grown-up scientists are just as prone to wishful thinking as little boys like Calvin. For example, centuries ago it was thought that insects and other small animals arose directly from spoiled food. This was easy to believe, because small animals were thought to be very simple (before the invention of the microscope, naturalists thought that insects had no internal organs). But as biology progressed and careful experiments showed that protected food did not breed life, the theory of spontaneous generation retreated to the limits beyond which science could not detect what was really happening. In the nineteenth century that meant the cell. When beer, milk, or urine were allowed to sit for several days in the containers, even closed ones, they always became cloudy from something growing in them. The microscopes of the eighteenth and nineteenth centuries showed that the growth was very small, apparently living cells. So it seemed reasonable that simple living organisms could arise spontaneously from liquids.

The key to persuading people was the portrayal of the cells as "simple." One of the chief advocates of the theory of spontaneous generation during the middle of the nineteenth century was Ernst Haeckel, a great admirer of Darwin and an eager popularizer of Darwin's theory. From the limited view of cells that microscopes provided, Haeckel believed that a cell was a "simple little lump of albuminous combination of carbon," not much different from a piece of microscopic Jell-O. So it seemed to Haeckel that such simple life, with no internal organs, could be produced easily from inanimate material. Now, of course, we know better.[18]

Likewise, in an age of scientific enlightenment we should know better than to believe in the evolutionary dogma that presupposes something comes from nothing. To hold to the law of energy conservation is to jettison cartoon characters for the canon—which clearly communicates that since God finished the work of creating (Gen. 2:1–3), He has been sustaining (conserving) all things by His power (Heb. 1:3).

Entropy

While the law of energy conservation is a blow to the theory of evolution, the law of entropy is a bullet to its head. Not only is the universe dying of heat loss, but according to entropy—also known as the second law of thermodynamics—everything runs inexorably from order to disorder and from complexity to decay. The theory of biological evolution directly contradicts the law of entropy in that it describes a universe in which things run from chaos to complexity and order. In evolution, atoms allegedly self-produce amino acids, amino acids auto-organize amoebas, amoebas turn into apes, and apes evolve into astronauts.

Mathematician and physicist Sir Arthur Eddington demonstrated that exactly the opposite is true: The energy of the universe irreversibly flows from hot to cold bodies.[19] The sun burns up billions of tons of hydrogen each second, stars burn out, and species eventually become extinct.

"If your theory is found to be against the second law of thermodynamics [entropy] I can give you no hope; there is nothing for it but to collapse in deepest humiliation"—Sir Arthur Eddington

While I would fight for a person's right to have faith in science fiction, we must resist evolutionists who attempt to brainwash people into thinking that evolution is science.[20] Evolution requires constant violations of the second law of thermodynamics in order to be plausible. In the words of Eddington, "If your theory is found to be against the second law of thermodynamics I can give you no hope; there is nothing for it but to collapse in deepest humiliation."[21]

Rather than humbling themselves in light of the law of entropy, evolutionists dogmatically attempt to discredit or dismiss it. First, they contend that the law cannot be invoked because it merely deals with energy relationships of matter, while evolution deals with complex life-forms arising from simpler ones. This, of course, is patently false. As a case in point, contemporary information theory deals with information entropy and militates against evolution on a genetic level.[22] While in an energy conversion system entropy dictates that energy will decay, in an informational system entropy dictates that information will become distorted.[23] As noted in *Scientific American,* "It is certain that the conceptual connection between information and the second law of thermodynamics is now firmly established."[24]

Furthermore, it is boldly asserted that entropy does not prevent evolution on earth since this planet is an open system that receives energy from the sun. This, of course, is nonsense. The sun's rays never produce an upswing in complexity without teleonomy (the ordering principle of life). In other words, energy from the sun does not produce an orderly structure of growth and development without information and an engine.[25] If the sun beats down on a dead plant, it does not produce growth but rather speeds up decay. If, on the other hand, the sun beats down on a living plant, it produces a temporary increase in complexity and growth. In the *Origins* film series, Dr. A. E. Wilder-Smith explains that the difference between a dead stick and a live orchid is that the orchid has teleonomy, which is information that makes the live orchid an energy-capturing and order-increasing machine.[26]

Finally, it has been suggested by some evolutionists—who appear to suffer from a sort of cognitive dissonance—that the law of entropy might not have operated in the distant past. Creationist Dr. Henry Morris points out that

> this assumption would be tantamount to the denial of the basic assumption of evolutionism, namely, that *present* processes suffice to account for the origin of things. In effect, this device would acknowledge the validity of the creationist approach, acknowledging that special creative processes operating only in the past are necessary to explain the world of the present.[27]

Thousands of years before empirical science formally codified the law of entropy, Scripture clearly communicated it. The prophet Isaiah and King David both declared that the heavens and the earth would "wear out like a garment" (Isa. 51:6; Ps. 102:25–26). Likewise, in the first century Paul the Apostle looked forward to the day when "the creation itself will be liberated from its *bondage to decay*" (Rom. 8:21; emphasis added).

In summary it should be noted that philosophical naturalism—the worldview undergirding evolutionism—can provide only three explanations for the existence of the universe in which we live. (1) The universe is merely an illusion. This notion carries little weight in an age of scientific enlightenment. As has been well said, "Even the full-blown solipsist looks both ways before crossing the street." (2) The universe sprang from nothing. As previously demonstrated, this proposition flies in the face of both the laws of cause and effect and energy conservation. There simply are no free lunches. The conditions that hold true in this universe prevent any possibility of matter springing out of nothing.[28] (3) The universe eternally existed. This hypothesis is devastated by the law of entropy, which predicts that a universe that has eternally existed would have died an "eternity ago" of heat loss.[29]

There is, however, one other possibility. It is found in the first chapter of the first book of the Bible: "In the beginning God created the heavens and the earth." In an age of empirical science, nothing could be more certain, clear, or correct.

We have used FACE as an acronym to help you remember salient points to respond to the farce of evolution. You should now be equipped to demonstrate that

the **F**ossil record is an embarrassment to evolutionists;

that **A**pe-men are fiction, fraud, and fantasy;

that **C**hance renders evolution not just improbable but also impossible; and

that **E**mpirical science supports the creation model for origins and militates against the theory of evolution. But there's more.

By adding the letter **R** we can change the acronym FACE into F-A-R-C-E. In this way you will be reminded of one more nail in the evolutionary coffin—namely, **r**ecapitulation.

During development, the fertilized egg progresses over thirty-eight weeks through what is, in fact, a rapid passage through evolutionary history: From a single primordial cell, the conceptus progresses through being something of a protozoan, a fish, a reptile, a bird, a primate and ultimately a human being. There is a difference of opinion among scientists about the time during a pregnancy when a human being can be said to emerge. But there is general agreement that this does not happen until after the end of the first trimester.

—Elie A. Schneour

Recapitulation provided a convenient focus for the pervasive racism of white scientists: they looked to the activities of their own children for comparison with normal, adult behavior in lower races.

—Stephen Jay Gould

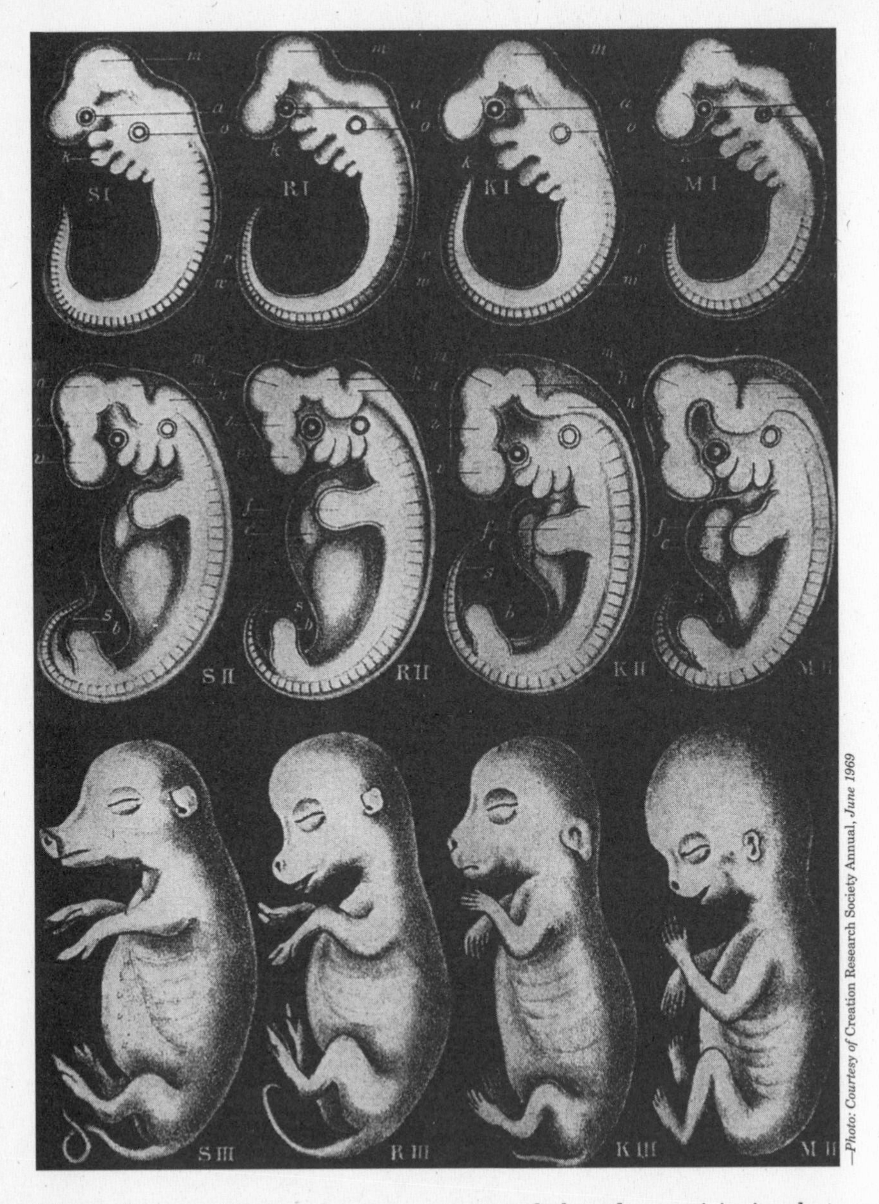

—*Photo: Courtesy of* Creation Research Society Annual, *June 1969*

Haeckel's recapitulation drawings are not only based on revisionism but have been used as justification for Roe v. Wade and for racism. Distorted charts intended to emphasize similarities in embryo development among (from left to right) a pig, bull, rabbit, and human.
—From Haeckel's *Anthropogeny*

CHAPTER 6

Recapitulation

Recapitulation, better known by the once popular evolutionary phrase "Ontogeny recapitulates phylogeny," is the notion that in the course of an embryo's development (ontogeny), the embryo repeats (recapitulates) the evolutionary history of its species (phylogeny).[1] Thus, at various points, an emerging human is a fish, a frog, and finally a fetus. This theory, first championed by a German biologist named Ernst Haeckel, is not only based on *revisionism* but has also been used as justification for *Roe v. Wade* and for *racism*.

Revisionism

In *Ontogeny and Phylogeny* Harvard professor Stephen Jay Gould points out that German scientist Wilhelm His exposed such "shocking dishonesty" on the part of Ernst

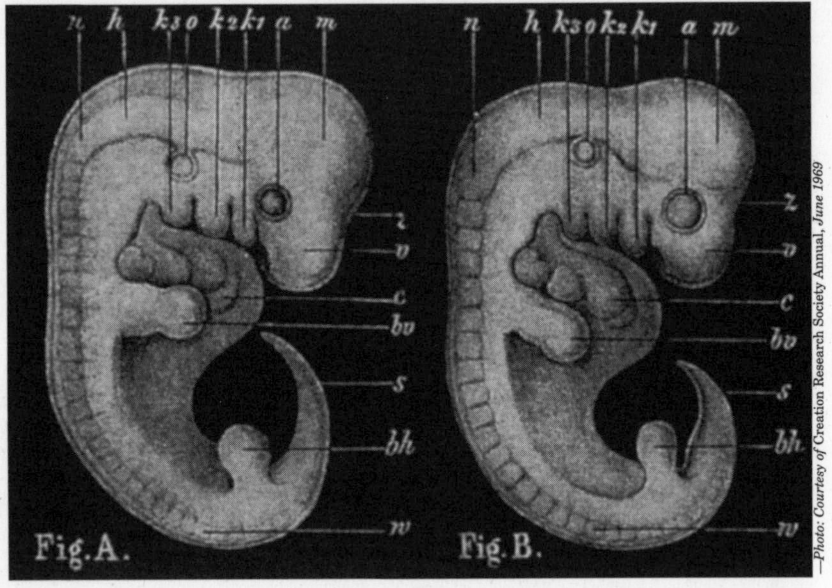

—Photo: Courtesy of Creation Research Society Annual, June 1969

Haeckel utilized deceptive data as well as doctored drawings to delude devotees. These doctored drawings designed to demonstrate the similarity between dog (Fig. A.) and human (Fig. B.) embryos were originally featured in *Natural History of Creation* 1868.

Haeckel that it rendered him unworthy "to be counted as a peer in the company of earnest researchers."[2] Tragically, despite acknowledging that the recapitulation theory has been discredited, Gould proceeded to write an entire book to demonstrate that it is still "one of the great themes of evolutionary biology."[3]

Sir Gavin de Beer of the British Natural History Museum was more circumspect. He was quoted as saying, "Seldom has an assertion like that of Haeckel's 'theory of recapitulation,' facile, tidy, and plausible, widely accepted without critical examination, done so much harm to science."[4] Haeckel not only utilized deceptive data but also used doctored drawings to delude his devotees.[5] His dishonesty was so blatant that he was charged with fraud by five professors and convicted by a university court at Jena.[6] His forgeries were subsequently made public with the 1911 publication of *Haeckel's Frauds and Forgeries.*[7]

Today the "recapitulations" most commonly referred to by educators and evolutionists are the "gill slits" in the "fish stage" of human embryonic growth. Dr. Henry Morris notes several reasons why this supposed recapitulation is entirely superficial. First, the human embryo never at any time develops gill slits and therefore never goes through a "fish stage." Furthermore, a fetus does not have fins or any other fish structures. Finally, every stage in the development of an embryo plays a crucial role in embryonic growth. Thus, there are no redundant vestiges of former evolutionary phases.[8]

Although Haeckel's frauds and forgeries were exposed more than half a century ago, modern studies in molecular genetics have further demonstrated the utter absurdity of the recapitulation theory. The DNA for a fetus is not the DNA for a frog and the DNA for a frog is not the DNA for a fish. Rather the DNA of a fetus, frog, fish, or falcon, for that matter, is uniquely programmed for reproduction after its own kind.[9]

Incredibly, such facts have not stopped men like Carl Sagan from affirming recapitulation. Fully aware that it had "gone through various cycles of scholarly acceptance and rejection," Sagan wrote:

> Haeckel held that in its embryological development, an animal tends to repeat or recapitulate the sequence

> that its ancestors followed during their evolution. And indeed in human intrauterine development we run through stages very much like fish, reptiles, and non-primate mammals before we become recognizably human. The fish stage even has gill slits, which are absolutely useless for the embryo who is nourished via the umbilical cord, but a necessity for human embryology: since gills were vital to our ancestors, we run through a gill stage in becoming human.[10]

This idea, of course, is not science; it's science fiction. For more than a century it has been well known that what Sagan referred to as "gill slits" are in reality essential parts of human anatomy. Far from being useless evolutionary vestiges, they are axiomatic to the development of a human embryo.[11]

Roe v. Wade

In *The Dragons of Eden* Sagan stated that determining when a fetus becomes human "could play a major role in achieving an acceptable compromise in the abortion debate."[12] In his estimation the transition to human "would fall toward the end of the first trimester or near the beginning of the second trimester of pregnancy."[13]

Shortly before he died, I watched him reiterate this odd predilection. Using recapitulation as the pretext, he shamelessly defended the painful killing of innocent human beings. Without so much as blushing, he communicated his contention that a first-trimester abortion does not constitute the painful killing of a human fetus but merely the termination of a fish or frog. Thus in Sagan's world, *Roe v.*

—Photo: UPI/CORBIS-BETTMANN

Popular evolutionist Dr. Carl Sagan used recapitulation as the pretext for communicating that a first trimester abortion does not constitute the painful killing of a human *fetus* but rather the termination of a *fish* or a *frog*.

Wade provided the legal framework for the slaughter of multiplied millions of creatures rather than children.

Sagan was not alone in his rationalization for abortion. His appeal to the science fiction of recapitulation is common fare. Curtin Winsor, Jr., the ambassador to Costa Rica, provides a classic case in point: In response to "After Roe," an article in the *National Review* by Father Richard John Neuhaus, Ambassador Winsor—with sarcasm dripping from his pen—wrote,

> If our well-meaning preacher has studied fetal anatomy, he might recall that the developing fetus actually retraces the evolutionary process. In a period of a few weeks, it begins as a single-cell creature and grows to be a fish (with gills), an amphibian, a reptile, and a

> mammal with a tail. Yes, at all times, this fetus has human potential, but that potential is implicit, not explicit, until around the twelfth week of pregnancy, when it manifests its human reality.[14]

Ambassador Winsor goes on to point out that recapitulation is a "reasonable basis upon which critical distinctions can be made in law." He specifically names "the law deriving from *Roe v. Wade,*" which "gives the unwilling mother the benefit of the doubt, at least until her fetus is explicitly human."[15] A similar example of incomprehensible ignorance in an age of scientific enlightenment is found in the words of Elie Schneour, director of the Biosystems Research Institute in La Jolla, California, and chairman of the Southern California Skeptics Society:

> *Ontogeny recapitulates phylogeny.* This is a fundamental tenet of modern biology that derives from evolutionary theory, and is thus anathema to creationism as well as to those opposed to freedom of choice. Ontogeny is the name for the process of development of a fertilized egg into a fully formed and mature living organism. Phylogeny, on the other hand, is the history of the evolution of a species, in this case the human being. During development, the fertilized egg progresses over thirty-eight weeks through what is, in fact, a rapid passage through evolutionary history: from a single primordial cell, the conceptus progresses though being something of a protozoan, a fish, a reptile, a bird, a primate and ultimately a human being. There is a difference of

> opinion among scientists about the time during a pregnancy when a human being can be said to emerge. But there is general agreement that this does not happen until after the end of the first trimester.[16]

Schneour, of course, is dead wrong; the notion that there is general agreement among scientists that a human being does not emerge during a pregnancy until after the end of the first trimester is merely the figment of a rather fertile imagination. While an emerging embryo may not have a fully developed personality, it does have full personhood from the moment of conception. French geneticist Jerome L. LeJeune bore eloquent testimony to this truth while testifying to a United States Senate subcommittee: "To accept the fact that after fertilization has taken place a new human has come into being is no longer a matter of taste or opinion. The human nature of the human being from conception to old age is not a metaphysical contention, it is plain experimental evidence."[17]

Dr. Hymie Gordon, professor of medical genetics and a physician at the prestigious Mayo Clinic, best summarized the perspective of science as follows:

> I think we can now also say that the question of the beginning of life—when life begins—is no longer a question for theological or philosophical dispute. It is an established scientific fact. Theologians and philosophers may go on to debate the meaning of life or purpose of life, but it is an established fact that all life, including human life, begins at the moment of conception.[18]

Recapitulation, of course, is just one of many arguments that have been used as justification for the atrocities that have resulted from *Roe v. Wade.* In light of both scientific and scriptural evidence that abortion is the painful killing of an innocent human being, I've developed the acronym A-B-O-R-T-I-O-N as a memorable tool to equip you to deal effectively with these arguments. This resource can be found in Appendix D. Annihilating Abortion Arguments.

Racism

Roe v. Wade is not the only ghastly consequence of the recapitulation theory; racism is another. This point is amplified by none other than Stephen Jay Gould, who notes that recapitulation served as a basis for Dr. Down in labeling Down Syndrome as "'Mongoloid idiocy' because he thought it represented a 'throwback' to the 'Mongolian stage' in human evolution."[19]

Tragically, as Gould points out, the term "'Mongoloid' was first applied to mentally defective people because it was then commonly believed that the Mongoloid race had not yet evolved to the status of the Caucasian race."[20] Dr. Henry Morris underscores the horror of this racist notion by noting that recapitulationists not only believe that human embryos recapitulate the evolutionary history of their ancestors but some like "Haeckel (and his disciple Adolph Hitler) used it to justify the myth of the Aryan super-race, destined to subjugate or obliterate other races."[21] In their view a "Caucasian human infant had to develop through stages corresponding to the 'lower' human races (hence, the origin

—Photo: AP/WIDE WORLD PHOTOS

Dr. Down believed Down's Syndrome represented a throwback to the Mongolian stage of human evolution—to wit his use of the term "Mongoloid idiocy."

of the term 'mongolism') before becoming a full-fledged member of the 'master' race."[22]

It is incredible to think that in light of such evidence, evolutionists such as Winsor, an American ambassador; Sagan, a successful scientist; and Schneour, the chairman of a skeptics society, would champion recapitulation. It is equally amazing to realize that Henry Fairfield Osborn, the leading American paleontologist of the first half of the twentieth century, had the temerity to say that "the Negroid stock is even more ancient than the Caucasian and Mongolian. . . . The standard of intelligence of the average adult Negro is similar to that of the eleven-year-old youth of the species *Homo sapiens.*"[23]

Creationists can only be grateful that Gould has not followed suit. In an article titled "Dr. Down's Syndrome," he "decries recapitulation's responsibility for the racism of the post-Darwinian era."[24] In his words, "Recapitulation provided a convenient focus for the pervasive racism of white scientists; they looked to the activities of their own children for comparison with normal, adult behavior in lower races."[25]

Sadly, although Gould has abandoned recapitulation, he has not abandoned evolution. Instead he pitifully attempts to prop up the crumbling edifice of evolution with novel notions such as punctuated equilibrium. It's one thing to blame Gould for evolution, it's quite another to blame God.

Theistic Evolution

Under the banner of "theistic evolution" a growing number of Christians maintain that God used evolution as His method for creation. This, in my estimation, is the worst of all possibilities. It is one thing to believe in evolution, it is quite another to blame God for it. Not only is theistic evolution a contradiction in terms—like the phrase *flaming snowflakes*—but as we have seen, it is also the cruelest, most inefficient system for creation imaginable. As Jacques Monod put it:

> [Natural] selection is the blindest, and most cruel way of evolving new species. . . . The struggle for life

> and elimination of the weakest is a horrible process, against which our whole modern ethic revolts. . . . I am surprised that a Christian would defend the idea that this is the process which God more or less set up in order to have evolution.[26]

An omnipotent, omniscient God does not have to painfully plod through millions of mistakes, misfits, and mutations in order to have fellowship with humans. Rather He can create humans in a microsecond. If theistic evolution is true, Genesis is at best an allegory and at worst a farce. And if Genesis is an allegory or a farce, the rest of the Bible becomes irrelevant. If Adam did not eat the forbidden fruit and fall into a life of constant sin terminated by death, there is no need for redemption.[27]

G. Richard Bozarth was right: "Christianity is—must be—totally committed to special creation as described in Genesis, and Christianity must fight with its full might, fair or foul, against the theory of evolution."[28] He was wrong, however, in saying that Christianity was fighting for its life. Quite the opposite is true. *Evolutionism* is fighting for its life. Rather than prop it up with theories like theistic evolution, Christians must be on the vanguard of demonstrating its demise.

Darwinism today is in much the same condition as Marxism was before its collapse. Its terminal condition cannot be successfully treated with medieval medications such as pseudosaurs or punctuated equilibrium. As the Soviet Union collapsed before our very eyes, so, too, the propped up corpse of evolution is ready for its final fall. As mathematician Dr. David Berlinski eloquently satirized, "Darwin's theory of evolution is the last of the great nineteenth-century mystery religions. And as we speak it is now following Freudianism and Marxism into the Nether regions, and I'm quite sure that Freud, Marx and Darwin are commiserating one with the other in the dark dungeon where discarded gods gather."[29]

While insiders in the evolutionary community are aware of their theory's desperate condition, the general public is as yet in the dark. That's precisely where you and I come in. We have the inestimable privilege to share the news that nothing could be more compelling in an age of scientific enlightenment than: "In the beginning God created the heavens and the earth."

Since the creation of the world God's invisible qualities—his eternal power and divine nature—have been clearly seen, being understood from what has been made, so that men are without excuse.

—the apostle Paul

—Photo: Victoria & Albert Museum, London/Art Resource, NY

Communicating the Good News.
Raphael (1483-1520), *Paul Preaching at Athens.*

Epilogue

It is not enough to use FACE as an acronym to remind us how to demonstrate the FARCE of evolution. It is crucial that we are also equipped to use our well-reasoned answers as a springboard or an opportunity for communicating that the God who created us also desires fellowship with us. While we cannot change anyone's heart—only the Holy Spirit can do that—we can equip ourselves to sensitively and effectively communicate our faith. The good news of the gospel should be such a part of us that presenting becomes second nature.*

The goal of winning the creation-evolution debate is not to demonstrate our mental acumen but rather to use our well-reasoned answers as opportunities for sharing the good news that God who created us desires fellowship with us as

*For a memorable program that will equip you to sensitively and effectively communicate your faith, write Hank Hanegraaff, Box 80250, Rancho Santa Margarita, CA 92688-0250 or call (949) 589-1504. Ask for *Personal Witness Training: Your Handle on the Great Commission.*

well. Thus, it is crucial that we are prepared to communicate that Christianity is not merely about religion—rather, it is about the truth through which human beings can have a relationship with the Creator of the Cosmos.

Relationship

The distinction between religion and relationship makes all the difference in the world. Religion is merely people's attempt to reach up and become acceptable to God through their own efforts—by living a good life, attempting to obey the Ten Commandments, or following the Golden Rule. Some religions even teach that this cannot be accomplished in one lifetime. Thus, people are reincarnated over and over until they finally become one with nirvana or one with the universe.

The problem with the answer provided by *religion* is that the Bible* says that in order to become acceptable to God, we must be absolutely *perfect!* As Jesus taught in His Sermon on the Mount—one of the most famous literary masterpieces in the history of humanity—"Be perfect, therefore, as your heavenly Father is perfect" (Matt. 5:48). Obviously no one is perfect; therefore, if we are ever going to know God here and now as well as rule and reign with Him throughout the eons of time, there has to be another way. And that way is found in a *relationship*.

*A typical response by a skeptic at this point is, "Who says that what the Bible says is true?" In response it is crucial to be able to demonstrate that the Bible is divine rather than human in origin. In fact, if you can demonstrate that the Bible was inspired by God rather than conspired by humans, you can answer a host of other objections by referring directly to Scripture. For an overview on how to demonstrate the inspiration of Scripture, turn to Appendix B. The Bible: Divine or Human in Origin?

—*Photo: Scala/Art Resource, NY*

Relationship
Michelangelo (1475–1564), *Creation of Adam.*

This is what the Christian faith is all about. It is not primarily a set of do's and don'ts. It's a personal relationship with God. That relationship does not depend on our ability to reach up and touch God through our own good works, but rather on God's willingness to reach down and touch us through His love.

By way of illustration, if I wanted to have a relationship with an ant, the only way I could do so would be to become one. Obviously I can't become an ant, but God did become a man. The Bible says that God in the person of Jesus Christ became flesh and lived for a while among us (John 1:14). He came into time and space to restore a relationship with His people that was severed by sin.

As you continue, it is crucial that you underscore the problem of sin. If people do not recognize that they are sinners, neither will they realize their need for a Savior.

Sin

Sin is not just murder, rape, or robbery. Sin is also failing to do the things we should and doing those things that we should not. In short, *sin* is a word to describe anything that fails to meet God's standard of perfection. Thus, sin is the barrier between us and a satisfying relationship with God. As Scripture puts it, "Your iniquities [sins] have separated you from your God" (Isa. 59:2).

Just as light and dark cannot exist together, neither can God and sin. And each day we are separated farther from God as we add to the account of our sin. But that's not the only problem. Sin also separates us from others. You need only read the newspaper or listen to a news report to see how true this really is.

—Photo: Foto Marburg / Art Resource, NY

Separation from God; Separation from one another.
Rubens (1577-1640), *Elevation of the Cross.*

Locally we read of murder, robbery, and fraud. Nationally we hear of corruption in politics, racial tension, and an escalating suicide rate. Internationally there are constantly wars and rumors of war. We live in a time when terrorism abounds and when the world as we know it can be instantly obliterated by nuclear aggression.

All of these things are symbolic of sin. The Bible says that we "all have sinned and fall short of the glory of God" (Rom. 3:23).[1] There are no exceptions to the rule. The problem is further compounded when we begin to understand who God is. Virtually every heresy begins with a misconception of the nature of God.

God

God is the perfect Father. We all have earthly fathers, but no matter how good—or bad, as the case may be—none are perfect. God, however, is the perfect Father. And as the perfect Father, He desires an intimate personal relationship with each of us. In His word He says, "I have loved you with an everlasting love" (Jer. 31:3).

Yet the same Bible that tells us of God's love also tells us that He is the perfect Judge. As such, God is absolutely just, righteous, and holy. The Bible says of God, "Your eyes are too pure to look on evil; you cannot tolerate wrong" (Hab. 1:13).

Herein lies the dilemma. On the one hand we see that God is the perfect Father who loves us. On the other hand, He is the perfect Judge whose very nature is too pure to tolerate our sin.

This dilemma is brought into sharper focus by a story I heard many years ago about a young man caught driving

—Photo: Alinari/Art Resource, NY

Perfect Father, Perfect Judge.
Baciccio (1639-1709), *Triumph of the Name of Jesus.*

under the influence after having committed several crimes. He was brought before a judge nicknamed the "hanging judge." Although the judge's integrity was beyond question, he always handed out the stiffest penalty allowable by law. It turns out that the judge was the young man's father. As you can imagine, everyone in the courthouse that day waited with bated breath to see how the judge would treat his own son. Would he show him favoritism as a father, or would he, as always, hand out the stiffest penalty allowable by law?

As the spellbound courtroom full of spectators looked on, the judge, without hesitation, issued the maximum penalty allowable by law. Then he took off his judicial robes, walked over to where his son stood, and paid the penalty his son could not pay. In that one act, he satisfied the justice of the law yet demonstrated his extraordinary love.

That, however, is but a faint glimpse of what God the Father did for us through His Son, Jesus Christ. You see, Jesus Christ—God Himself—came to earth to be our Savior and to be our Lord.

It is important to be prepared to point out that through His resurrection, Jesus demonstrated that He does not stand in a line of peers with Buddha, Mohammed, or any other founder of a world religion. They died and are still dead, but Christ had the power to lay down His life and to take it up again.*

Jesus Christ

As our Savior, Jesus lived the perfect life we cannot live. Earlier I pointed out that Scripture says we need to be per-

*For an overview on how to demonstrate the fact of the resurrection, see Appendix C. The Greatest FEAT in the Annals of Recorded History.

fect to be acceptable to God. Well, Jesus Christ came into time and space to be perfection for us. "God made him [Jesus Christ] who had no sin to be sin for us, so that in him we might become the righteousness of God" (2 Cor. 5:21).

This is the great exchange over which all of the Bible was written. God took our sins and placed them on Jesus Christ, who suffered and died to pay the debt we could not pay. Then, wonder of wonders, He gave us the perfect life of Jesus Christ. He took our sin and gave us His perfection as an absolutely free gift. We cannot earn it or deserve it; we can only live a life of gratitude for this gift. Jesus not only died to be our Savior, He also lives to be our Lord.

As our Lord, Jesus Christ gives our lives meaning, purpose, and fulfillment. This is a particularly exciting thought when we stop to realize that the One who wants to be our Lord is the very One who spoke and caused the universe to leap into existence. He not only made this universe and everything in it, but He also made *us*. He knows all about us. He loves us and wants us to have a satisfying life here and now, and an eternity of joy with Him in heaven forever.

The Bible says, "If you confess with your mouth, 'Jesus is Lord,' and believe in your heart that God raised him from the dead, you will be saved" (Rom. 10:9). The resurrection of Jesus Christ is an undeniable fact of history. And that's not just anyone's opinion. That was the opinion of Dr. Simon Greenleaf, who was the greatest authority on legal evidence in the nineteenth century. He was also the famous Royall Professor of Law at Harvard and was directly responsible for Harvard's rise to eminence among American law schools.

After examining the evidence for the resurrection of Jesus Christ, he said, in effect, that it was the most well-attested

—Photo: Scala/Art Resource, NY

Our Savior, Our Lord.
Rouault (1871–1958). Ecce Homo (Behold the Man).

fact of ancient history. Through the undeniable fact of the resurrection, Jesus demonstrated that He was God in human flesh. To receive Jesus Christ as Savior and Lord we need only take two steps. The one step is to *repent,* the other to *receive.*

To those who are touched by our communication of the gospel, we must also communicate the two steps necessary to experience a genuine relationship with God.

Two Steps

The first step involves repentance. *Repentance* is an old English word that describes a willingness to turn from sin toward Jesus Christ. It literally means a complete U-turn on the road of life—a change of heart and a change of mind. It means a willingness to follow Jesus Christ and receive Him as Savior and Lord. In the words of Christ, "The time has come. . . . The kingdom of God is near. Repent and believe the good news!" (Mark 1:15).

The second step to demonstrate true belief is a willingness to receive. To truly receive is to trust in and depend on Jesus Christ alone to be the Lord of our lives here and now and our Savior for all eternity.

It takes more than knowledge (the devil knows about Jesus and trembles). It takes more than agreeing that the knowledge is accurate (the devil agrees that Jesus is Lord). True saving faith entails not only knowledge and agreement, but also trust. By way of illustration, when you are sick, you can know a particular medicine can cure you. You can even agree that it's cured thousands of others. But until you trust it enough to take it, it cannot cure you. In like manner, you can know about Jesus Christ, and you can

—Photo: Foto Marburg/Art Resource, NY

Repent and Receive.
Gruenewald (1455–1528). *Mary Magdalene.*

agree that He has saved others, but until you personally place your trust in Him, you will not be saved. The requirements for eternal life are not based on what *you can do* but on what *Jesus Christ has done.* He stands ready to exchange His perfection for your imperfection.

To those who have never received Him as Savior and Lord, Jesus says, "Here I am! I stand at the door and knock. If anyone hears my voice and opens the door, I will come in" (Rev. 3:20).[2] Jesus knocks on the door of the human heart, and the question He asks is, Are you ready *now* to receive Me as Savior and Lord?

According to Jesus Christ, those who repent and receive Him are "born again"—not physically, but spiritually (John 3:3). And with this spiritual birth must come spiritual growth.

Growth

First, no relationship can flourish without constant, heartfelt communication. This is true not only in human relationships, but also in our relationship with God. If we are to nurture a strong walk with our Savior, we must be in constant communication with Him. The way to do that is through prayer.

You do not need a special vocabulary to pray. You can simply speak to God as you would to your best friend. The more time you spend with God in prayer, the more intimate your relationship will be. And remember, there is no problem, great or small, that He cannot handle. If it's important to you, it's important to Him.

In addition to praying, it is crucial to direct new believers to God's revelation of Himself—the Bible. The Bible not only forms the foundation of an effective prayer life, it is foundational to every other aspect of Christian living as well. While prayer is our primary way of communicating with God, the Bible is God's primary way of communicating with us. Nothing should take precedence over getting into the Word and getting the Word into us.

If we fail to eat well-balanced meals on a regular basis, we will eventually suffer the physical consequences. What is true of our outer selves is also true of our inner selves. If we do not regularly feed on the Word of God, we will spiritually starve.

I generally recommend that new believers begin by reading one chapter from the Gospel of John each day. As they do, they will experience the joy of having God speak to them directly through His Word. Jesus said, "I am the bread of life. He who comes to me will never go hungry, and he who believes in me will never be thirsty" (John 6:35).

Finally, it is crucial for new believers to become active participants in a healthy, well-balanced church. In Scripture the church is referred to as the body of Christ. Just as our body is one and yet has many parts, so the body of Christ is one but is composed of many members. Those who receive Christ as the Savior and Lord of their lives are already a part of the church universal. It is crucial, however, that they become vital, reproducing members of a local body of believers as well.

Scripture exhorts us "not [to] give up meeting together, as some are in the habit of doing" (Heb. 10:25). It is in the local church where God is worshiped through prayer, praise, and proclamation; where believers experience fellowship with one another; and where they are equipped to reach others through the testimony of their love, their lips, and their lives.

I began this epilogue by pointing out that Christianity is not merely a religion, but rather a relationship between humans and their Creator. The apostle Paul underscored this point in his sermon on Mars Hill. He stood in the midst of the Aeropagus and shouted,

> Men of Athens! I see that in every way you are very religious. For as I walked around and looked carefully at your objects of worship, I even found an altar with this inscription: TO AN UNKNOWN GOD. Now what you worship as something unknown I am going to proclaim to you.
>
> The God who made the world and everything in it is the Lord of heaven and earth and does not live in temples built by hands. And he is not served by human hands, as if he needed anything, because he himself gives all men life and breath and everything else. From one man he made every nation of men, that they should inhabit the whole earth; and he determined the times set for them and the exact places where they should live. God did this so that men would seek him and . . . find him, though he is not far from each one of us. "For in him we live and move and have our being." As some of your own poets have said, "We are his offspring."
>
> Therefore since we are God's offspring, we should not think that the divine being is like gold or silver or stone—an image made by man's design and skill. In the past God overlooked such ignorance but now he commands all people everywhere to repent. For he has set a day when he will judge the world with

justice by the man he has appointed. He has given proof of this to all men by raising him from the dead. (Acts 17:22–31)

Some sneered at the message. Others were saved.

APPENDIX A

Death Moves

Jim McLean, one of the premiere teachers in the game of golf, coined the phrase *death moves*. As he explains, "A Death Move is an action or position that is so far from the ideal that it will cause you to hit poor shots forever."[1] Thus when he sees someone making a death move, he doesn't merely recommend a modification, he recommends an immediate and complete change.

As there are "death moves" in the game of golf, so, too, there are "death moves" in the field of apologetics or the defense of the Christian faith. When a death move is made in golf, the result is that the golf ball does not move toward the intended target. In sharing and defending the good news of the gospel, the far more serious result is that the message inevitably misses its mark.

Death Move #1—Bad Arguments

A classic example of a bad argument is the Darwin legend. In order to demonstrate the falsity of evolution, Bible-believing Christians for more than a century have passed on the story of Darwin's deathbed conversion.

In *The Darwin Legend* James Moore describes "sawdust trail revivalists" gloating, "Have you heard about that old reprobate, Charles Darwin? Why, on his deathbed he renounced his theory of evolution and was wondrously saved!"[2] Likewise, contemporary counterfeit revivalist John Arnott says that "Darwin later renounced his theory of evolution and, as a born-again Christian, died in peace in his retirement home in southern India."[3]

On the other side of the ledger are evolutionists who attempt to counter revivalists by loudly protesting that Darwin "knew to the last that religion is a fraud and that science has discovered the real creator!"[4]

First it should be noted that whether Darwin did or did not renounce evolution does *not* speak to the issue of whether evolution is true or false. Maybe Darwin renounced evolution because he was senile or he had taken a mind-altering drug. He may have even just hedged his bets with some "eternal fire insurance."

Furthermore, as followers of the One who proclaimed Himself to be not only the way and the life but also the truth, we must set the standard for the evolutionist, *not* vice versa. James Fegan was correct in calling the Darwin legend "an illustration of the recklessness with which the Protestant Controversialists seek to support any cause they are advocating."[5]

Finally, in *The Darwin Legend* Moore painstakingly documents the fact that there is *no* substantial evidence that Darwin ever repented, but there is abundant evidence that he consistently held to his evolutionary paradigm.

Death Move #2—Trying to Talk Someone into the Kingdom

A common death move happens when we believe that someone can be talked into the kingdom of God. While this death move may originate from a sincere desire to reach the lost, the consequences are often devastating.

No matter how eloquent we may be, we cannot change anyone else's heart—only the Holy Spirit can change the heart. Thus, while it is our responsibility to "always be prepared to give an answer to everyone who asks [us] to give the reason for the hope that [we] have" (1 Peter 3:15), it is God who changes the heart.

Ultimately it is not that others cannot believe, it is that they will not believe. In other words, it is often not a matter of the mind but a matter of the will. As Jesus Christ declared: "This is the verdict: Light has come into the world, but men loved darkness instead of light because their deeds were evil. Everyone who does evil hates the light, and will not come into the light for fear that his deeds will be exposed" (John 3:19–20).

The Christian faith is reasonable; however, reason alone will not compel people to embrace Christ. By nature our hearts are set in opposition to God. Unless He supernaturally intervenes, lasting change will not take place. Thankfully, in His providence, God ordains both the ends and the means. While He uses our well-reasoned answers, ultimately He Himself changes the heart.

I am utterly convinced that if we are "prepared to give an answer," God will bring into our paths those whose hearts He has prepared. Rather than having to run around frantically grabbing people by the lapels, it is our responsibility to prepare ourselves to be the most effective tools in the hands of Almighty God.

Death Move #3—"Winging It"

A common tendency for Christians who do not know how to answer a question is to simply "wing it." More often than not they are motivated by the notion that the lack of a *ready* answer may result in the compromise of their witness. In reality, quite the opposite is true—the lack of a *right* answer may compromise their witness. The proper response is to do the necessary research and then return with the correct information.

As a case in point, imagine that as a student you are attempting to convince a biology teacher that the creation model for origins is valid. As you point out that there are no transitions from one species to another in the fossil record, the teacher brings up *Archaeopteryx* as a transitional form between reptiles and birds. Defensively you blurt out, "*Archaeopteryx* is really just a hoax."

Unfortunately, at this point your witness has been compromised. While *Archaeopteryx* is clearly not a transition from reptiles to birds, it is just as clearly not a hoax. Rather than speak out of ignorance, it is far better to admit that you haven't researched *Archaeopteryx* but you will, and return with a reasoned argument. As we discovered in the section on pseudosaurs, the truth about *Archaeopteryx* will demonstrate to anyone with an open mind that it is a better testimony to special creation than to evolution.

APPENDIX B

The Bible: Human or Divine?

To defend the faith we must be equipped to demonstrate that the Bible is divine rather than human in origin. If we can successfully accomplish this, we can answer a host of other objections simply by appealing to Scripture. To chart our course I will use the acronym M-A-P-S. Since most Bibles have maps in the back, this should prove to be a memorable association.

M = Manuscripts. Since we don't have the original biblical manuscripts, we must ask, "How good are the copies?" The answer is that the Bible has stronger manuscript support than any other work of classical literature, including those of Homer, Plato, Aristotle, Caesar, and Tacitus. The reliability of Scripture is also confirmed through the eyewitness credentials of the authors. Moses, for example, participated in and was an eyewitness to the remarkable events of the Egyptian captivity, the exodus, the forty years in the desert, and Israel's final encampment before entering the Promised Land, all of which are accurately chronicled in the Old Testament.

The New Testament has the same kind of eyewitness authenticity. Luke says that he gathered the eyewitness testimony and "carefully investigated everything" (Luke 1:1–3). Peter reminded his readers that the disciples "did not follow cleverly invented stories" but "were eyewitnesses of [Jesus'] majesty" (2 Peter 1:16).

Secular historians—including Josephus (before A.D. 100), the Roman Tacitus (around A.D. 120), the Roman Suetonius (A.D. 110), and the Roman governor Pliny the Younger (A.D. 110)—confirm the many events, people, places, and customs chronicled in the New Testament. Early church leaders such as Irenaeus, Tertullian, Julius Africanus, and Clement of Rome—all writing before A.D. 250—also shed light on New Testament historical accuracy. Even skeptical historians agree that the New Testament is a remarkable historical document.

A = Archaeology. Over and over again, comprehensive fieldwork (archaeology) and careful biblical interpretation affirm the reliability of the Bible. It is telling when a secular scholar must revise biblical criticism in light of solid archaeological evidence.

For years, critics dismissed the book of Daniel, partly because there was no evidence that a king named Belshazzar ruled in Babylon during that period. Later archaeological research, however, confirmed that the reigning monarch, Nabonidus, appointed Belshazzar as his coregent while he was waging war away from Babylon.

One of the best-known New Testament examples concerns the books of Luke and Acts. A biblical skeptic, Sir

William Ramsay, was trained as an archaeologist and then set out to disprove the historical reliability of this portion of the New Testament. But through his painstaking Mediterranean archaeological trips, he became converted as, one after another, the historical allusions of Luke were proved accurate. Truly, with every turn of the archaeologist's spade, we continue to see evidence for the trustworthiness of Scripture.

P = Prophecy. The Bible records predictions of events that could not have been known or predicted by chance or common sense. Surprisingly, the predictive nature of many Bible passages was once a popular argument against the reliability of the Bible. Critics argued that various passages were written later than the biblical texts indicated because they recounted events that happened sometimes hundreds of years later than when they supposedly were written. They concluded that, subsequent to the events, literary editors went back and "doctored" the original, nonpredictive texts.

But this is simply wrong. Careful research *affirms* the predictive accuracy of the Bible. For example, the book of Daniel (written before 530 B.C.) accurately predicts the progression of kingdoms from Babylon through the Medo-Persian Empire, the Greek Empire, and then the Roman Empire, culminating in the persecution and suffering of the Jews under Antiochus IV Epiphanes, his desecration of the temple, his untimely death, and freedom for the Jews under Judas Maccabeus (165 B.C.).

Old Testament prophecies concerning the Phoenician city of Tyre were fulfilled in ancient times, including prophecies that the city would be opposed by many nations

(Ezek. 26:3); its walls would be destroyed and towers broken down (26:4); and its stones, timbers, and debris would be thrown into the water (26:12). Similar prophecies were fulfilled concerning Sidon (Ezek. 28:23; Isa. 23; Jer. 27:3–6; 47:4) and Babylon (Jer. 50:13, 39; 51:26, 42, 43, 58; Isa. 13:20, 21).

Since Christ is the culminating theme of the Old Testament and the Living Word of the New Testament, it should not surprise us that prophecies regarding Him outnumber all others. Many of these prophecies would have been impossible for Jesus to deliberately conspire to fulfill—such as His descent from Abraham, Isaac, and Jacob (Gen. 12:3; 17:19); His birth in Bethlehem (Mic. 5:2); His crucifixion with criminals (Isa. 53:12); the piercing of His hands and feet on the cross (Ps. 22:16); the soldiers' gambling for His clothes (Ps. 22:18); the piercing of His side and the fact that His bones were not broken at His death (Zech. 12:10; Ps. 34:20); and His burial among the rich (Isa. 53:9). Jesus also predicted His own death and resurrection (John 2:19–22). Predictive prophecy is a principle of Bible reliability that often reaches even the hard-boiled skeptic!

S = Statistics. It is statistically preposterous that any or all of the Bible's specific, detailed prophecies could have been fulfilled through chance, good guessing, or deliberate deceit. When you look at some of the improbable prophecies of the Old and New Testaments, it seems incredible that skeptics—knowing the authenticity and historicity of the texts—could reject the statistical verdict: The Bible is the Word of God, and Jesus Christ is the Son of God, just as Scripture predicted many times and in many ways.

The Bible was written over a span of 1,600 years by forty authors using three languages (Hebrew, Aramaic, and

Greek) on hundreds of subjects. And yet there is one consistent, noncontradictory theme that runs through it all: God's *redemption of humankind.* Clearly, statistical probability concerning biblical prophecy is a powerful indicator of the trustworthiness of Scripture.

The next time someone denies the reliability of Scripture, just remember the acronym MAPS, and you will be equipped to give an answer and a reason for the hope that lies within you. Manuscripts, archaeology, prophecy, and statistics not only chart a secure course through the turnpikes of skepticism but also demonstrate conclusively that the Bible is indeed divine rather than human in origin.

APPENDIX C

The Greatest FEAT in History

The resurrection of Jesus Christ is the greatest F-E-A-T in the annals of history. Through the resurrection, Jesus demonstrated that He does not stand in a line of peers with Buddha, Mohammed, or any other founder of a world religion. They died and are still dead, but Christ is risen. As someone has well said, the resurrection is the very capstone in the arch of Christianity; if it is removed, all else crumbles. It is the singular doctrine that elevates Christianity above all the pagan religions of the ancient Mediterranean world.

"If Christ has not been raised, your faith is futile; you are still in your sins. . . . If only for this life we have hope in Christ, we are to be pitied more than all men" (1 Cor. 15:17, 19). It is precisely because of the resurrection's strategic importance that each Christian must be prepared to defend its historicity. Let's use the acronym FEAT as a reminder to help us do just that.

F = Fact. The resurrection of Jesus Christ is an undeniable fact of history. And that's not just anyone's opinion. That was the opinion of Dr. Simon Greenleaf, who was the greatest authority on legal evidence of the nineteenth century. He was also the famous Royall Professor of Law at Harvard and was directly responsible for the school's rise to eminence among American law schools. After being goaded by his students into examining the evidence for the resurrection, Greenleaf suggested that any cross-examination of the eyewitness testimonies recorded in Scripture would result in "an undoubting conviction of their integrity, ability and truth." In 1846 Dr. Greenleaf wrote a defense for the resurrection titled *An Examination of the Testimony of the Four Evangelists by the Rules of Evidence Administered in the Court of Justice.*

E = Empty Tomb. The first major support for the resurrection of Christ is the empty tomb. Even the enemies of Christ had to admit that the tomb was empty. The record shows that they went so far as to attempt to bribe the guards to say the body had been stolen (Matt. 28:11–15). If the Jewish leaders had stolen the body, they could have later openly displayed it to prove that Jesus had not risen from the dead. Although many flawed theories have been concocted over the years, the fact of the empty tomb has never been refuted.

A = Appearances. The second major support for the resurrection is the appearance of Christ after the resurrection. He appeared to more than five hundred witnesses at a single time (1 Cor. 15:6). He also appeared to many other people, providing "many convincing proofs" of His resurrection (Acts 1:3). Christ in His resurrection body was even touched on two occasions (Matt. 28:9; John 20:17), and He challenged the disciples and Thomas to feel His wounds (Luke 24:39; John 20:27).

T = Transformation. The third great apologetic for the resurrection is the radical transformation that took place in the lives of Christ's disciples. Before the resurrection, they might best have been characterized as cowards. After the resurrection, they were transformed into lions of faith. Despite intense persecution and the threat of cruel deaths, they testified to the truth of the resurrection. While it is conceivable that some might die for what they believed to be the truth, it is inconceivable that so many would die for what they knew to be false. As Greenleaf put it, "If it were morally possible for them to have been deceived in this matter, every human motive operated to lead them to discover and avow their error. . . . If then their testimony was not true, there was no possible motive for this fabrication."[1]

Not only did the resurrection of Christ transform the disciples from cowards into lions of faith, but His resurrection also continues to transform lives today. Because Christ lives, the Scripture says, we will live also. In an instant, in the twinkling of an eye, our bodies shall be transformed into resurrected bodies like unto His resurrected body. Indeed, the evidence for Christ's resurrection is so overwhelming that no one can examine it with an open mind without becoming convinced of its truth.

APPENDIX D

Annihilating Abortion Arguments

They sacrificed their sons and daughters in the fire . . . and sold themselves to do evil in the eyes of the LORD, provoking him to anger" (2 Kings 17:17).

For hundreds of years the Lord had warned the Israelites through His prophets. Now it was too late! Darkness had descended upon the Promised Land. The people of Israel had become the slaves of the mighty Assyrians. Although the tribe of Judah to the south had miraculously survived the initial onslaught, they somehow blithely managed to ignore the lesson of history.

Second Kings tells us that Ahaz, king of Judah, "walked in the ways of the kings of Israel and even sacrificed his son in the fire, following the detestable ways of the nations the LORD had driven out before the Israelites" (16:3).

The nation of Israel had indeed become a mirror reflection of the pagan culture that surrounded it. True prophets continued to warn God's people that their wickedness would inexorably lead to destruction, but their words fell on deaf ears. The rulers of the land had become so corrupt that they even hired false prophets to tell them what their itching ears wanted to hear.

Finally, the inevitable occurred. The ax of God's judgment fell. Babylon leveled Jerusalem, and the people of Judah were driven from their land of promise.

Today America, like ancient Israel, is turning a deaf ear to the lesson of history. We have repeatedly violated God's commands, as if we could do so with impunity. We have failed to heed the warnings of His prophets and have embraced the new paganism of our times. Indeed, our ways have become detestable to the Lord; we have forgotten His command:

> When you enter the land the LORD your God is giving you, do not learn to imitate the detestable ways of the nations there. *Let no one be found among you who sacrifices his son or daughter in the fire,* who practices divination or sorcery, interprets omens, engages in witchcraft, or casts spells, or who is a medium or spiritist or who consults the dead. Anyone who does these things is detestable to the LORD, and because of these detestable practices the LORD your God will drive out those nations before you. You must be blameless before the LORD your God. (Deut. 18:9–13; emphasis added)

Christian philosopher Francis Schaeffer warned us that abortion would be the watershed issue of our era. He said, "Of all the subjects relating to the erosion of the sanctity of human life, abortion is the keystone. It is the first and crucial issue that has been overwhelming in changing attitudes toward the value of life in general."[1]

Schaeffer's warning has tragically fallen on deaf ears. For more than two decades we have sacrificed our children on the altars of hedonism. And even now, the ax of God's judgment has been laid to the root.

Two thousand years ago Christ warned us that "the time will come when you will say, 'Blessed are the barren women, the wombs that never bore and the breasts that never nursed!'" (Luke 23:29). The present-day abortion holocaust has driven those words home in dramatic fashion. Consider the statements of some of the leading spiritual and secular leaders of our age:

- Beverly Harrison (professor of Christian ethics at Union Theological Seminary): *"Infanticide is not a great wrong.* I do not want to be construed as condemning women who, under certain circumstances, quietly put their infants to death" (emphasis in original).[2]

- Esther Langston (professor of social work at the University of Nevada, Las Vegas): "What we are saying is that abortion becomes one of the choices and the person has the right to choose whatever it is that is . . . best for them in the situation in which they find themselves, be it abortion, to keep the baby, to adopt it, to sell it, to leave it in a dumpster, to put it on your porch, whatever; it's the person's right to choose."[3]

- Mary S. Calderone, M.D. (head of the Sex Information and Education Council of the United States): "We have yet to beat our drums for birth control in the way we beat them for polio vaccine, we are still unable to put babies in the class of dangerous epidemics, even though this is the exact truth."[4]

- Margaret Sanger (the late founder of Planned Parenthood): "The most merciful thing a large family can do for one of its infant members is to kill it."[5]

- Nobel prize laureate James Watson (codiscoverer of DNA): "Because of the limitations of present detection methods, most birth defects are not discovered until birth. . . . However if a child was not declared alive until three days after birth . . . the doctor could allow the child to die if the parents so choose and save a lot of misery and suffering."[6]

Perhaps most frightening of all, President Clinton signed into law the National Institute of Health Revitalization Act. As a direct result, it is now legal not only to abort but also to carve up murdered babies and use them for fetal tissue for research.[7]

While pondering this horrifying reality, remember that the present-day holocaust is government funded. That means you and I are footing the bill![8]

Make no mistake: "Choice" advocates like Clinton, Congress, and the courts are not the friends of children. America's unthinking submission to their twisted arguments is moving us progressively toward social genocide of a magnitude eclipsing that of Hitler, Stalin, Somalia, and the Serb-Croate conflict.

The movement's own label—"pro-choice"—is a twisted deception, covering up a nationally sanctioned holocaust in which the "right" to choose to kill a child reigns supreme over

- the baby's human rights.
- the rights of the parents of a pregnant minor.
- the rights of the baby's father.
- the mother's right to accurate information about fetal development and the negative consequences of abortion.
- the rights of society to protect all its members—no matter what their social status, economic situation, or physical limitations.

What Is Abortion?

Those who continue to fight legislation restricting abortion are in reality *not* "pro-choice." Rather, they are singularly "pro-murder." While the rhetoric has served to camouflage the carnage, abortion is really the painful killing of an innocent human being.

Painful

Abortion is painful because the methods employed to kill a preborn child involve burning, smothering, dismembering, and crushing. Dr. James Dobson offers a terrifying description of one method of abortion called Dilation and Extraction (D & X):

> Over two days the cervix is dilated. Then an ultrasound device and forceps are used to reach in and grab the baby's feet. The little body is pulled downward until just the head remains in the cervix. Next

the abortionist grasps the nape of the neck and cuts open the back of the skull with blunt scissors. A device called a cannula is then inserted into the wound and the brain material is sucked out. If kidneys or other organs are desired, they are removed while the child is still partially in the vagina. Initially at least, these surgical procedures are performed on a live baby who has not specifically been anesthetized (although the mother's medication may reduce some of the pain).[9]

Abortion is also performed by a procedure called Dilation and Curettage (D & C), in which a tiny hoe is used to chop the baby's body to pieces. The placenta is then scraped off the wall of the uterus and the body is subsequently reassembled to ensure that nothing has been left behind. Other methods include:

- Saline solution: A salt solution is injected into the amniotic fluid, burning the skin off the baby who, after thrashing in the uterus for a number of hours, is reduced to a shriveled corpse.
- Suction: Presently two-thirds of all abortions in the United States and Canada are carried out using a suction tube, which tears the child apart and deposits the pieces into a jar.
- Hysterotomy: This is similar to a Cesarean section, except it is designed for the express purpose of killing rather than saving the baby.
- Prostaglandin: This chemical is injected into the uterine muscle, causing it to react violently, thus expelling

the child (the few children who avoid decapitation resulting from the violent contractions are exterminated after delivery).

Killing

Abortion involves killing; the zygote, which fulfills the criteria needed to establish the existence of biological life (including metabolism, development, the ability to react to stimuli, and cell reproduction), is indeed terminated.

Innocent

While it is true that everyone is conceived and born in sin, preborn children are innocent because they have done nothing wrong. They deserve protection, not capital punishment.

Human Being

The living baby in the mother's womb is a human being; the child is the product of human parents and has a totally distinct human genetic code. This truth that abortion terminates the life of a human being is substantiated by science:

- As Dr. Micheline Matthew-Roth, a principal research associate at Harvard Medical School's Department of Medicine, puts it, "It is scientifically correct to say that an individual human life begins at conception, when egg and sperm join to form the zygote, and this developing human always is a member of our species in all stages of its life."[10]
- French geneticist Jerome L. LeJeune bore eloquent testimony to the truth of Dr. Matthew-Roth's remarks

when he gave the following sworn statement to a United States Senate subcommittee: "To accept the fact that after fertilization has taken place a new human has come into being is no longer a matter of taste or opinion. The human nature of the human being from conception to old age is not a metaphysical contention, it is plain experimental evidence."[11]

- Perhaps Dr. Hymie Gordon, professor of medical genetics and a physician at the prestigious Mayo Clinic, best summarized the perspective of science when he said, "I think we can now also say that the question of the beginning of life—when life begins—is no longer a question for theological or philosophical dispute. It is an established scientific fact. Theologians and philosophers may go on to debate the meaning of life or purpose of life, but it is an established fact that all life, including human life, begins at the moment of conception."[12]

Long before science substantiated the truth that abortion is the painful killing of an innocent human being, the psalmist summarized the view of sacred Scripture with these words:

> For you created my inmost being;
> you knit me together in my mother's womb.
> I praise you because I am fearfully and wonderfully made;
> your works are wonderful,
> I know that full well.
> My frame was not hidden from you
> when I was made in the secret place.

When I was woven together in the depths of the earth,
your eyes saw my unformed body.
All the days ordained for me
were written in your book
before one of them came to be. (Ps. 139:13–16)

In light of the fact that both science and Scripture corroborate the view that abortion is the painful killing of an innocent human being, it is incumbent upon Christians to do everything in our power to halt the spread of this enormous evil. There are indeed many fronts on which our battle must be waged. Ultimately, however, lasting change only comes when the hearts of people are transformed. For when the heart is transformed, a person's behavior is revolutionized as well. Because of the transcendent importance of this issue, I've developed the acronym A-B-O-R-T-I-O-N to serve as a memorable tool to help believers annihilate abortion arguments.

Remember, however, the goal is not to win an argument but rather to use well-reasoned answers to the arguments of abortion advocates as springboards or opportunities to share a message of life and light.

A = Ad Hominem: Attacking people rather than arguing principles. *Ad hominem* arguments are designed to distract attention from the *real* issue—namely, that abortion is the killing of an innocent human being. Comedienne Whoopi Goldberg used this tactic when she suggested that abortion rights advocates would take pro-lifers more seriously if they were willing to adopt babies slated for abortion.[13]

What this *ad hominem* argument is really saying is, "If you won't adopt my babies, don't tell me I can't kill them!"

That, of course, makes as much sense as forbidding me from intervening when I see my neighbor sexually abusing a child unless I am willing to adopt that child.

The "adoption argument" completely evades the basic morality or immorality of abortion. Instead, it is an attempt to attack character in order to avoid the case against abortion.

Another common *ad hominem* attack involves the media portrayal of pro-lifers as wild-eyed fanatics. For instance, the death of abortionist Dr. David Gunn has been widely used to stereotype those who believe in the sanctity of life as "social terrorists." Senator Edward M. Kennedy has gone so far as to say, "Attacks on clinics are not isolated incidents and health care providers are living in fear for their lives. . . . No doctors should be forced to go to work in a bullet-proof vest."[14] Senator Barbara Boxer exudes, "American women have seen their doctors' offices transformed from safety zones into war zones."[15]

A final *ad hominem* attack worth mentioning is the fallacy that pro-lifers are inconsistent because they denounce abortion while supporting capital punishment. In fact, many pro-lifers do *not* support capital punishment. But for the many others that do, this argument still fails on many counts. The most obvious rebuttal is that abortion involves the killing of an innocent human being while capital punishment involves the killing of someone who has been found guilty of a capital crime.

B = Biblical Pretexts: Using biblical texts out of context as a pretext for abortion. Pro-abortionists typically use biblical pretexts to retain some semblance of religiosity while at the same time espousing the radical planks of the pro-abortion movement. The most common argument in this area is that Scripture nowhere specifically condemns abortion or identifies it as the killing of an innocent human being. Such an argument, however, obscures the fact that the Bible depicts preborn children as living beings who are fully human. (See, for example, Ps. 139:13–16.) Furthermore, Scripture clearly denounces the killing of an innocent human being as murder. Thus, abortion is a violation of the Sixth Commandment (Exod. 20:13).

Ironically, one of the most commonly used biblical pretexts for abortion is found only one chapter after God's explicit command "Thou shall not murder":

> If men struggle with each other and strike a woman with child so that she has a miscarriage, yet there is no further injury, he shall surely be fined. . . . But if there is any further injury, then you shall appoint as a penalty life for life, eye for eye, tooth for tooth, hand for hand, foot for foot, burn for burn, wound for wound, bruise for bruise. (Exod. 21:22–25 NASB)

The argument goes something like this: If a man strikes a pregnant woman and causes her to have a spontaneous abortion, the penalty is merely a fine. However, if the woman dies, the penalty is death. Thus, no life was taken, according to Exodus 21, unless the woman died.

Thus interpreted, this passage is not being used but abused to support abortion. Let's take a closer look at what the Hebrew text (as correctly translated in the NIV) really says:

> If men who are fighting hit a pregnant woman and she gives birth prematurely but there is no serious

> injury [the implication here is that *no* death is involved], the offender must be fined whatever the woman's husband demands and the court allows. But if there is serious injury, you are to take life for life [in other words, if the woman *or* child should die, the appropriate punishment is death].

Another biblical pretext, typically referred to as the "argument from breath," involves Genesis 2:7: "The LORD God formed the man from the dust of the ground and breathed into his nostrils the breath of life, and the man became a living being."

The argument from breath is frequently presented in the following manner: God did not consider Adam to be a "living soul" until He had breathed the "breath of life" into him. Thus a child does not become a human being until he or she begins to breathe.

Dispensing with this argument is a simple matter. Adam was inanimate before God breathed the breath of life into him. Conversely, as science demonstrates, the conceptus or preborn child is alive from the very moment of conception. It is important to note that the breath of life exists in the preborn child from the moment of conception. In reality, it is the form, not the fact, of oxygen transfer (breath) that changes at birth.

O = Opium: As opium dulls the senses chemically, so the term-twisting tactics of pro-abortionists deaden the perception of the human carnage caused by abortion. In 1844, Karl Marx wrote, "Religion . . . is the opium of the people."[16] While history has demonstrated that true religion doesn't deaden the senses but rather brings life, it may well be said that the terminology of pro-abortionists is specifically designed to mentally dull the senses of an unquestioning public. For example, a pro-abortion stance is called pro-choice; babies are demoted to the status of POCs or products of conception; killing unwanted children is repositioned as exercising freedom of choice; and committed pro-lifers are tagged as political extremists or even social terrorists.

The list of camouflaged terms employed by pro-abortionists seems endless. Unless we learn to unmask the language of the pro-abortion lobby, millions will continue to become morally numb on the opium of clever code words.

R = Rape and Incest: Used as an emotional appeal designed to deflect serious consideration of the pro-life platform. Rape and incest are the hard-case "what-ifs" pro-abortionists raise in almost every public forum: "How can you deny a hurting young girl safe medical care and freedom from the terror of rape or incest by forcing her to maintain a pregnancy resulting from the cruel and criminal invasion of her body?" The emotion of this argument often deflects serious examination of its merits and is commonly used as a pretext for abortion on demand.

It is important to note that the incidence of pregnancy as a result of rape is extremely small (one study put it at 0.6 percent).[17] As philosopher Francis Beckwith astutely points out, "To argue for abortion on demand from the hard cases of rape and incest is like trying to argue for the elimination of traffic laws from the fact that one might have to violate some of them in rare instances, such as when one's spouse or child needs to be rushed to the hospital."[18] If we had legislation restricting abortion for all reasons *other than* rape or incest, we would save the vast majority of the

1.8 million preborn babies who die annually in America through abortion.

Furthermore, one does not obviate the real pain of rape or incest by compounding it with the murder of a child; two wrongs obviously do not make a right. The very thing that makes rape evil also makes abortion evil. In both cases, an innocent human being is brutally dehumanized. The real question that must be answered is whether preborn children are indeed fully human. As has already been documented, the answer is a resounding yes!

T = Toleration: The "great commandment" of the pro-abortion movement. Perhaps the most common argument pro-abortionists level against opponents is the argument from toleration. For example: "We're not making *you* have an abortion, so why can't you be tolerant of those who choose to?" Translated: "Don't impose your antiquated morals on me." At first blush this argument may seem reasonable, but upon closer examination its inherent weakness becomes readily apparent. Imagine applying this line of reasoning to the issue of rape by saying, "Don't like rape? Don't rape anyone. Just don't impose your morality on me!"

This false standard of tolerance is frequently supported by an appeal to religious pluralism. In this context, pro-abortionists argue that government should not take one theory of life and impose it on others. The obvious problem with this argument is that not only is the pro-abortion position forced on Christians, but Christians are required to fund it as well. Incredibly, pro-abortionists fail to perceive their own violation of this standard: They're intolerant of those who think tolerance is less important than preserving innocent human lives!

Every society has the obligation to impose morals on its citizens. Toleration works in the world of expressing opinions, not in a crowded movie theater when someone chooses to yell "Fire!" We may be tolerant of one's religious views, but not if they include enslaving grandmothers or cannibalizing teenagers.

Separation between church and state does not extend to divorcing all moral values from the state. If this were the case, we would need to eliminate all legislation that has anything in common with a religious point of view—including the very idea of social law itself.

Remember, tolerance when it comes to personal relationships is a virtue, but tolerance when it comes to truth is a travesty.

I = Inequality: Women "forced" to have babies cannot compete successfully with their male counterparts. Inequality between the sexes is one of the most bizarre arguments put forth by the pro-abortion movement. "Women who are forced to be pregnant," it is said, "can't compete in employment with men and so cannot be truly equal unless they have an escape from unwanted pregnancy." This is like saying, "Women can't be equal to men without reconstructive surgery." How much more sexist can an argument become?

Imagine, however, applying this standard to children outside the womb. Following this "logic," women should be permitted to abandon children who pose a threat to their opportunities for advancement.

Another form of the "inequality argument" is graphically portrayed through the image of a rusty coat hanger. Prior to *Roe v. Wade,* pro-abortionists claimed that because of financial inequality, women who could not afford to fly to another country to get an abortion were condemned to performing abortions on themselves with rusty coat hangers. To add credibility to this assertion, statistics ranging from 5,000 to 10,000 deaths per year due to illegal abortions continue to be widely circulated.[19]

Dr. Bernard Nathanson, a former leader of the National Abortion Rights Action League (NARAL), had this to say about these preposterous statistics: "I confess I knew the *figures were totally false,* and I suppose the others did too. . . . But in the 'morality' of the revolution, it was a useful figure" (emphasis added).[20]

According to the U.S. Bureau of Vital Statistics, the true figure of the women who died from illegal abortions in 1972—the year prior to *Roe v. Wade*—is thirty-nine. It is also questionable whether any of these thirty-nine women died as a result of using a coat hanger. As unpleasant as it may be, consider for a moment the dexterity needed to dislodge a conceptus from the uterine wall using a crude tool like a coat hanger. The truth of the matter is that the pro-abortion argument from inequality is not only illogical, but deliberately deceptive as well.

O = Operation Rescue: The number-one straw-man argument of the pro-abortion lobby. Operation Rescue has been condemned for using the same lines of argumentation and social protest popularized by the civil rights movement—a movement pro-abortion advocates usually extol. Furthermore, Operation Rescue has been grossly misrepresented, presumably to dismiss all pro-life activities as "extremist." The truth, however, is that just as abolitionists harbored escaped slaves in defiance of the laws before the

Civil War, compassionate Europeans hid Jews from the legally sanctioned extermination of the Nazis, and civil rights marchers violated segregation laws, so Operation Rescue members believe their nonviolent, peaceful interventions to protect preborn children are obeying God (see Acts 4:19). Nonetheless, it needs to be recognized that many of the mainstream pro-life groups do not approve of using nonviolent civil disobedience.

While it might be argued that the tactics of Operation Rescue are not the most effective means of stemming the tide of abortion, it is patently false to caricature members of Operation Rescue as social terrorists or worse. Any unbiased evaluation of the principles and procedures embraced by the leadership of this organization must conclude that they have consistently advocated *nonviolent civil disobedience.* It is therefore wrong for pro-abortionists to attempt to tie Operation Rescue and pro-lifers generally to the few tragic instances in which pro-life extremists have resorted to violence and murder.

On a personal note, I am grateful to God for the documented evidence of lives that have been saved through the self-sacrifice of dedicated men, women, and children involved in this movement.

N = Nonpersonhood: The emerging embryo may not have a fully developed personality, but it does have complete personhood. Nonpersonhood is perhaps the trickiest of the contemporary pro-abortion arguments. Pro-abortionists once argued that the preborn baby was not fully human. Now, however, advances in science have forced most people to concede that the "product of conception" is truly human.

As a result, a new version of this argument goes something like this: "The preborn child may be a human life, but it does *not* possess personhood."

Dr. Francis Beckwith exploded the latest version of this myth when he wrote,

> From a strictly scientific point of view, there is no doubt that the development of an individual human life begins at conception. Consequently, it is vital that the reader understand that she did not come from a zygote, she once was a zygote; she did not come from an embryo, she once was an embryo; she did not come from a fetus, she once was a fetus; she did not come from an adolescent, she once was an adolescent.[21]

The abortion epidemic ravaging America today is the tragic consequence of a decadent society that no longer values the individual human worth of each member, that worships the idol of "Selfism," and that replaces the objective Word of God with subjective preferences and social mores.

One-third of the children conceived in America this year will be savagely slaughtered before they are born. Yet this horrifying holocaust can be halted if those who value human life, worship the true God, and obey His Word will become informed, committed, and involved.

APPENDIX E

Human Cloning

As has been well said, "The only thing necessary for evil to triumph is for good men to do nothing." The stark reality of this sentiment was borne out in 1973 when Christians quietly passed by a major battle in the war against abortion. Two and a half decades later, the far-reaching impact of that loss is being felt in a raging debate over human cloning. While Pandora's box is already open, we must do all that is permissible to speak out and prevent a human clone from emerging.

Richard Seed, who holds farcical metaphysical views along with a Ph.D. in physics, has refueled the debate on human cloning. His preposterous plan to produce 200,000 human clones per year has caught the attention of politicians and pastors, as well as the president of the United States. According to Seed, "Cloning and the reprogramming of DNA is the first serious step in becoming one with God."[1] Following Seed's outrageous remarks, President Clinton immediately pushed Congress for a five-year moratorium on human cloning.[2]

Clinton's actions, however, may have come too late. Laws granting constitutional protection regarding reproductive choices,[3] as well as laws giving women carte blanche over abortion decisions,[4] may well limit the right of the government to interfere with human cloning experimentation. "It is hard to imagine a judicial decision consistent with current Supreme Court interpretations of abortion law," syndicated columnist Linda Chavez points out, "that would uphold the right of the government to outlaw cloning."[5] In other words, the prevailing logic that permits a woman to terminate the life of a child in the womb may well equally apply to allowing cloning.

The cloning issue took on greater urgency in 1997 when a team of Scottish embryologists produced the first successful clone of an adult mammal—a sheep named Dolly. In the process of experimentation, 276 out of 277 cloning attempts failed.[6] This is an important consideration in light of Seed's proposals for cloning human beings. In theory, human cloning is similar to animal cloning.[7] Thus for Richard Seed to succeed in his biological schemes, he would have to harvest female eggs, fuse them with the nuclei of human cells, and then implant the reworked eggs in the wombs of surrogate mothers, dooming almost all of the embryos to failure and death. While Seed sees this as an intoxicating proposition, the leaders of nineteen European nations signed a treaty declaring that cloning is "contrary to human dignity and constitutes a misuse of biology and medicine."[8] Christian leaders must speak out clearly and concisely against human cloning as well.

First, it is important to point out that producing a human clone would of necessity require experimentation on hundreds if not thousands of live human embryos. In reality (since embryos are fully human), the entire process would be the moral equivalent of the human experiments carried out by Nazi scientists under Adolf Hitler.

Furthermore, it should be noted that the issues concerning cloning and abortion are inextricably woven together. For example, if defects were detected in developing clones, abortion would no doubt be the solution of choice. And make no mistake—that an embryo has full personhood from the moment of conception is no longer a theological proposition, it is plain scientific fact. French geneticist Jerome LeJeune explains, "To accept the fact that after fertilization has taken place a new human has come into being is no longer a matter of taste or opinion. The human nature of the human being from conception to old age is not a metaphysical contention, it is plain experimental evidence."[9]

Finally, it should be emphasized that cloning has serious implications regarding what constitutes a family. While children are the result of spousal reproduction, clones are essentially the result of scientific replication.[10] Which raises the question: Who owns the clone? It is terrifying to consider that the first human clone might well be "owned and operated" by none other than Richard Seed.

Notes

Before You Begin

1. I have adapted this humorous way of describing *Pithecanthropus erectus* from Phil Saint, *Fossils That Speak Out* (Greensboro, NC: Saint Ministries International, 1985), 42.

2. Those who once placed their faith in *Pithecanthropus erectus* have had to move on to newer ape-men fantasies. *Pithecanthropus erectus* itself has "evolved" into the newer classification: *Homo erectus.*

3. Louis Bounoure, Director of Research at the French National Center for Scientific Research, Director of the Zoological Museum and former president of the Biological Society at Strasbourg. As quoted in

Advocate (March 1984): 17. As quoted in Paul S. Taylor, *The Illustrated Origins Answer Book* 4th ed. (Mesa, AZ: Eden Productions, 1993), 116.

Charting the Course

1. F. Darwin, ed., *The Life and Letters of Charles Darwin,* vol. 1 (London: John Murray, 1888), 45. As quoted in Michael Denton, *Evolution: A Theory in Crisis* (Bethesda, MD: Adler & Adler, 1986), 25. The notion that Darwin was ever a Bible-believing creationist is widely disputed. In fact, his grandfather Erasmus—the real inventor of the theory of evolution—was an eighteenth century rationalist.

2. Denton, 25.

1. Truth or Consequences

1. Sir Julian Huxley, *Essays of a Humanist* (New York: Harper & Row, 1964), 125. As quoted in John Ankerberg and John Weldon, *Darwin's Leap of Faith* (Eugene, OR: Harvest House, 1998), 39.

2. Ernst Mayr, "The Nature of the Darwinian Revolution," *Science* (2 June 1972): 981. As quoted in Henry M. Morris, *The Long War Against God* (Grand Rapids, MI: Baker Books, 1989), 20.

3. Michael Denton, *Evolution: A Theory in Crisis* (Bethesda, MD: Adler & Adler, 1986), 15.

4. Ibid., 358.

5. Ibid. It is important to note that Denton argues that Darwinism is "the only truly scientific theory of evolution." He says, "Reject Darwinism and there is, in effect, no scientific theory of evolution" (355).

6. Henry M. Morris, *The Long War Against God* (Grand Rapids, MI: Baker Books, 1989), 18.

7. G. Richard Bozarth, "The Meaning of Evolution" in *American Atheist* (February 1978): 19, 30.

8. Cf. Morris, *The Long War Against God,* 72, 74. The tragic fact that the twentieth century has witnessed more bloodshed than any other previous century nullifies an often-repeated accusation against Christianity, namely, that Christianity cannot be true because more suffering and death have resulted in the name of God than for any other reason. At the end of the twentieth century, however, with clear hindsight, we must not fail to recognize the ripened fruit of atheistic philosophies that have been fully actualized in this century's wars of extermination—in Marxist/Leninist totalitarianism, in Hitler's militaristic version, and so on (not including the horrendous holocaust of untold millions of babies through abortion). While it is tragically true that many people throughout history have misused

the name of God to justify wickedness, there is no question that at the end of the twentieth century the blood of multiplied millions drips from the hands of the atheistic, materialistic mentality. (Thanks to Phillip Johnson for emphasizing this point.)

9. Julian Huxley, Associated Press dispatch, Address at Darwin Centennial Convocation, Chicago University, 27 November 1959. See Sol Tax, ed., *Issues in Evolution* (Chicago: University of Chicago Press, 1960), 252. As quoted in Henry M. Morris, *That Their Words May Be Used Against Them* (El Cajon, CA: Institute for Creation Research, 1997), 111.

10. I first heard this story in a sermon by Dr. D. James Kennedy. Aldous Huxley, Sir Julian's brother, expressed similar sentiments in describing his own intellectual development: "Like so many of my contemporaries, I took it for granted that there was no meaning [to the world]. This was partly due to the fact that I shared the common belief that the scientific picture of an abstraction from reality was a true picture of reality as a whole; partly also to other, non-intellectual reasons. I had motives for not wanting the world to have a meaning; consequently assumed that it had none, and was able without any difficulty to find satisfying reasons for this assumption. . . .

    "For myself, as, no doubt, for most of my contemporaries, the philosophy of meaninglessness was essentially an instrument of liberation. The liberation we desired was simultaneously liberation from a certain political and economic system and liberation from a certain system of morality. We objected to the morality because it intefered with our sexual freedom. . . . The supporters of these systems claimed that in some way they embodied the meaning (a Christian meaning, they insisted) of the world. There was one admirably simple method of confuting these people and at the same time justifying ourselves in our political and erotic revolt: we could deny that the world had any meaning whatsoever. . . ." Aldous Huxley, *Ends and Means* (London: Chatto & Windus, 1938), 269–70, 273.

11. As of 1988, The Centers for Disease Control and Prevention estimates that the likelihood of new marriages in the United States ending in divorce is 43 percent (National Center for Health Statistics; http://www.cdc.gov/nchswww/fastats/divorce.html). The *Los Angeles Times* records, "Experts predict that nearly half of all new marriages will end with a judge parceling out houses and cars and children like so much chattel" (Sheryl Stolberg, "Signs Indicate Shift in Stance Toward Divorce," *Los Angeles Times,* 27 May 1996, A-1).

12. The Alan Guttmacher Institute (AGI), a research organization founded by Planned Parenthood, reports that more than 31 million legal abortions occurred in the United States between 1973 and 1994; AGI also reports that each year an estimated 50 million abortions occur worldwide (http://www.agi-usa.org/pubs-/fb_abortion2/fb_abort2.html). Concerning abortion, see Appendix D. Annihilating Abortion Arguments.

13. The Department of Veterans Affairs estimates a total of 1,089,200 deaths in American military service through America's wars (August 1997, www.va.gov-/pressre1/amwars97.html). The Joint United Nations Programme on HIV/AIDS estimates 11.7 million people worldwide have died from AIDS since the epidemic began (as reported by the Centers for Disease Control, 27 January 1998, www.cdc.gov/nchstp-/hiv_aids/stats/internat.html).

14. Kim Paintr, "Kiedis's Red-Hot Radio Spot," *USA Today,* 5 January 1994. As quoted in James Dobson, "Family News from Dr. James Dobson," February 1994, 4–5.

15. Letter from Charles Darwin to W. Graham, 3 July 1881, *Life and Letters of Charles Darwin,* vol. 1, 316, cited in Gertrude Himmelfarb, *Darwin and the Darwinian Revolution* (London: Chatto & Windus, 1959), 343. As quoted in Henry M. Morris, *Scientific Creationism,* public school edition (San Diego: C.L.P. Publishers, 1981), 179; emphasis added.

16. The quote continues:

    At the same time the anthropomorphous apes . . . will no doubt be exterminated. The break between man and his nearest allies will then be wider, for it will intervene between man in a more civilized state, as we may hope, even than the Caucasian, and some ape as low as a baboon, instead of as now between the Negro or Australian and the gorilla.

Charles Darwin, *The Descent of Man.* chap. VI "On the Affinities and Genealogy of Man," sect. "On the Birthplace and Antiquity of Man." As in Robert Maynard Hutchins, ed., *Great Books of the Western World,* vol. 49, *Darwin* (Chicago: Encyclopedia Britannica, 1952), 336.

17. It has been argued that Darwin, in the subtitle to *The Origin of Species by Means of Natural Selection,* probably does not intend, by the term *race,* to speak to the issue of distinctions among humans. However, his overall theory clearly extends the meaning of *race* to encompass the concept of *subspecies* distinctions among groups of humans. Thus, in his view, "the more civilized so-called Caucasian *races* have beaten the Turkish hollow in the struggle for existence" (Letter from Charles Darwin to W. Graham, 3 July 1881, *Life and Letters of Charles Darwin,* vol. 1, 316, cited in Gertrude Himmelfarb, *Darwin and the Darwinian Revolution* [London: Chatto & Windus, 1959], 343. As quoted in Morris, *Scientific Creationism,* 179; emphasis added).

18. *Agnostic,* coined from Greek, literally means "someone who lacks knowledge." The term was suggested by Huxley in 1869 to refer to one who thinks it is impossible to know whether there is a God, or a future life, or anything beyond material phenomena (*Webster's New Twentieth Century Dictionary of the English Language,* unabridged, 2d ed. [New York: Simon and Schuster, 1983], 37).

19. Thomas H. Huxley, *Lay Sermons, Addresses and Reviews* (New York: Appleton, 1871), 20. As quoted in Morris, *The Long War Against God,* 60; emphasis added.

20. Morris, *The Long War Against God,* 63. Ales Hrdlicka was one of the chief founders of institutional physical anthropology in the United States and was based at the American Museum of Natural History.

    Ernst Haeckel was a philosophical forerunner of Adolf Hitler in Germany and is most famous for his now debunked recapitulation theory (see chapter 6).

    E. A. Hooton was one of the primary founders of institutional physical anthropology in the United States and was based at Harvard.

21. It is a pervasive and consistent theme throughout Scripture that all people are created in the image of God and are of equal value (Gen. 1:27, 28; 9:6; James 3:9; cf. Eph. 4:24; Col. 3:10). "From one man [Adam] he made every nation of men, that they should inhabit the whole earth; and he determined the times set for them and the exact places where they should live" (Acts 17:26). Also, to parallel Galatians 3:28, Malachi 2:10 reads, "Have we not all one Father? Did not one God create us? Why do we profane the covenant of our fathers by breaking faith with one another?"

22. C. H. Woolston, "Jesus Loves the Little Children" (n.p., n.d.).

23. H. F. Osborn, "The Evolution of Human Races," *Natural History* (January/February 1926); reprinted in *Natural History* (April 1980): 129. As quoted in Morris, *The Long War Against God,* 62.

24. Marvin L. Lubenow, *Bones of Contention: A Creationist Assessment of the Human Fossils* (Grand Rapids, MI: Baker Books, 1992), 47.

25. Arthur Keith, *Evolution and Ethics* (New York: Putnam, 1947), 230. As quoted in Morris, *The Long War Against God,* 76.

26. Jacques Barzun, *Darwin, Marx, Wagner* (Garden City, NY: Doubleday, 1958), 8, cited in Morris, *The Long War Against God,* 83.

27. Daniel Goleman, "Lost Paper Shows Freud's Effort to Link Analysis and Evolution," *New York Times,* 10 February 1987, 19. As quoted in Morris, *The Long War Against God,* 33.

28. Ibid.

29. Darwin, *The Descent of Man,* 566.

30. While Scripture candidly acknowledges the existence of slavery, it never condones it. In the last book of the Bible, Revelation chapters 17 and 18, God pronounces final judgment on an evil world system that perpetuates slavery.

31. Denton, *Evolution: A Theory in Crisis,* 358.

2. Fossil Follies

1. Colin Patterson, personal letter to Luther Sunderland, 10 April 1979. As quoted in Luther D. Sunderland, *Darwin's Enigma,* 4th ed. (Santee, CA: Master Books, 1988), 89. In a later interview, Sunderland asked Patterson whether he knew of any good transitional forms. Patterson affirmed his prior statement that he did not know of any that he would try to support. Sunderland writes, "Throughout his interview he denied having transitional fossil candidates for each specific gap between the major different groups. He said that there are kinds of change in forms taken in isolation but there are none of these sequences that people like to build up" (p. 90).

   Patterson, of course, is a committed evolutionist. By "transitional form" I take him to mean a fossil that he can confidently say represents a form that lies between two fundamentally different species exhibiting wholly different structures and functions in an actual evolutionary line of descent—or that is directly ancestral to any such fundamentally different species. Furthermore, Patterson writes, "Fossils may tell us many things, but one thing they can never disclose is whether they were ancestors of anything else" (*Evolution* [London: British Museum of Natural History, 1978], 133. As quoted in W. R. Bird, *The Origin of Species Revisited,* vol. 1 [New York:

Philosophical Library, 1989], 183). Thus, while Patterson, along with most other evolutionists, does think that *Archaeopteryx,* for example, is evidence of some general evolutionary relationship between dinosaurs and birds (see "Pseudosaurs," p. 34), he thinks he cannot say with any confidence that *Archaeopteryx* is an intermediate, transitional form in the actual evolutionary line between dinosaurs and birds.

2. As Phillip Johnson has pointed out, the word "species" can be a bit of a trap. To evolutionists it can simply mean "isolated breeding groups," and in that sense trivial transitions can be said to have occurred. Of course, what we do not see are the sorts of transitions that would actually mean something and which evolutionary theory requires in order to stand: transitions between fundamentally different species that exhibit wholly different structures and functions (e.g., transitions between dinosaurs and birds, between apes and humans, etc.). (See also the distinction between microevolution and macroevolution in footnote 5 on pg. 172.)

3. The absence of verifiable transitions is striking in light of the fact that the theory of evolution utterly depends on the fossil record. Pierre-Paul Grasse, who held the prestigious Chair of Evolution at the Sorbonne for thirty years, wrote, "Naturalists must remember that the process of evolution is revealed only through fossil forms. A knowledge of paleontology

is, therefore, a prerequisite; only paleontology can provide them with the evidence and reveal its course or mechanisms" (*Evolution of Living Organisms* [New York: Academic Press, 1977], 4. As quoted in Henry M. Morris, *That Their Words May Be Used Against Them* [El Cajon, CA: Institute for Creation Research, 1997], 163).

4. David M. Raup, "Conflicts Between Darwin and Paleontology," *Bulletin, Field Museum of Natural History* (January 1979): 22, 25. As quoted in Paul S. Taylor, *The Illustrated Origins Answer Book,* 4th ed. (Mesa, AZ: Eden, 1993), 108; emphasis added.

5. The theory of biological evolution maintains that living things (plants, animals, humans, etc.) have descended with modification from shared, common ancestors.

   Macroevolution refers to large-scale changes—where one species transforms into another completely different species. For example, birds are said to have evolved from dinosaurs. This process would require the addition of new information to the genetic code.

   Microevolution refers to changes in the gene expressions of a given type of organism but does not produce a completely different species. For example, through selective breeding dogs ranging from Great Danes to Chihuahuas have been produced from wolves. This process, perhaps misnamed, does not require new information because the changes are a function of the genetic makeup already present in the gene pool of the species.

Gradualism refers to the theory that macroevolution proceeds through the slow and basically constant accumulation of many small changes in order to effect large changes. This theory predicts that the fossil record would provide abundant evidence of intermediary life forms as one species is progressively transformed into another.

6. Duane T. Gish, *Evolution: The Fossils Still Say No!* (El Cajon, CA: Institute for Creation Research, 1995), 130.

7. Ibid., 133, citing A. J. Charig, *A New Look At Dinosaurs* (London: Heinemann, 1979), 139.

8. Gish, 133 (see 133–39). Gish writes:

Since creatures within each family, order, or class are so highly variable, it would be predictable on the basis of the creation model that animals in different orders and classes would have some characteristics in common. Even humans share characteristics in common with reptiles. For example, we share in common the vertebrate eye. Among other characteristics, evolutionists emphasize that Archaeopteryx had teeth, a long tail, and claws on the wings, which, it is claimed, are reptilian characteristics, inherited from a reptilian ancestor. . . . [However,] Archaeopteryx did not have reptile-like teeth, but teeth that were uniquely bird-like, similar to teeth found in a number of other fossil birds . . . [having] unserrated teeth with constricted bases and expanded roots, while theropod dinosaurs, its alleged ancestors, had serrated teeth with straight roots. Furthermore, it should not be surprising that some birds had teeth, since this is true of all other vertebrates. Some fish have teeth, some do not. Some amphibians have teeth, some do not. Some reptiles have teeth, some do not. Most mammals have teeth, but some do not. . . . The long tail is supposed to be a reptilian feature, but, of course, some reptiles have short tails, while many have long tails. . . . A number of modern birds have claws on their wings. (p. 138)

Sunderland notes that "the tail bone and feather arrangement on swans are very similar to those of *Archaeopteryx*" (Luther D. Sunderland, *Darwin's Enigma,* 4th ed. [Santee, CA: Master Books, 1988], 74).

9. *The New Encyclopedia Britannica,* Micropaedia, 15th ed., vol. I (Chicago: Encyclopedia Britannica, 1981), 486. Gish, 133.

10. Gish, 133.

11. Ibid., 137, citing Tim Beardsley, *Nature* 322 (1986): 677; Richard Monastersky, *Science News* 140 (1991): 104–5; Alan Anderson, *Science* 253 (1991): 35.

12. *The New Encyclopedia Britannica,* 486. Gish, 132.

13. Gish, 133.

14. Allan Feduccia, *Science* 259 (1993): 790–93. As quoted in Gish, 135.

15. *The New Encyclopedia Britannica,* 486. Michael Denton, *Evolution: A Theory in Crisis* (Bethesda, MD: Adler & Adler 1986), 204–7. Gish, 134–35.

16. Feather anatomy description adapted from James F. Coppedge, *Evolution: Possible or Impossible?* (Northridge, CA: Probability Research in Molecular Biology, 1993), 215. Denton, 202.

17. Adapted from Denton, 203.

18. Pierre Lecomte du Nouy, *Human Destiny* (New York: Longmaus, Green and Co., 1947), 72. See also Gish, 140–41.

19. Stephen Jay Gould and Niles Eldridge, *Paleobiology* 3 (1977): 147. As quoted in Gish, 139.

20. Henry M. Morris and Gary E. Parker, *What Is Creation Science?,* rev. ed. (El Cajon, CA: Master Books, 1987), 138.

21. *American Scientist* (January/February 1979), cited in Morris and Parker, 138.

22. Comparing evolution to fairy tales was popularized by Duane T. Gish; see Gish, 5.

23. *Newsweek,* 3 November 1980. As quoted in Morris and Parker, 142.

24. If evolutionists begin with the presupposition that a

Creator had no role in the origin of life, then they will not be swayed in the least by the absence of evidence for evolution and the abundance of evidence for creation.

25. See Gish, 339–56.

26. Richard B. Goldschmidt, *The Material Basis of Evolution* (New Haven: Yale University Press, 1940), 395. As quoted in Gish, 344.

27. I first heard a version of this joke from D. James Kennedy in 1980.

28. Morris and Parker, 148. *The Wonderful Egg* was published in 1958 by Ipcar.

29. Stephen Jay Gould, *Natural History* vol. 86, no. 6 (1977): 22. As quoted in Gish, 341.

30. Gould (Harvard University), along with Niles Eldridge (American Museum of Natural History) and Stephen Stanley (Johns Hopkins University), has been the main proponent of punctuated equilibrium. As noted by Gish, "Gould, apparently embarrassed by his rather hasty and overenthusiastic support of the hopeful monster notion which he had voiced in his 1977 article, was attempting to extricate himself by denying that Goldschmidt really meant what he had said" (p. 346). Gish, however, documents that Goldschmidt surely did mean what he said (see Gish, 344–47).

31. Gould, *Natural History,* vol. 86, no. 5 (1977). As quoted in Gish, 346–47.

32. Gish, 354–55.

33. Ibid., 355.

34. Morris and Parker, 150. See also note 5, page 172–3.

35. Gould, *Natural History,* vol. 86, no. 5 (1977), 13. As quoted in Gish, 346. More recently, Gould has said that "paleontologists have discovered several superb examples of intermediary forms and sequences, more than enough to convince any fair-minded skeptic about the reality of life's physical genealogy" ("Hooking Leviathon by Its Past," *Natural History* [May 1994]: 8). Remember, however, that it was largely the "extreme rarity of transitional forms" that Gould intended to explain with his theory of punctuated equilibrium. If hopeful monsters are not possible and gradual changes are not abundantly represented in the fossil record, the "several . . . intermediary forms and sequences" (none of which have been conclusively identified) are at best mysterious anomalies that will convince only those who look at the fossil record with a prior commitment to evolutionary theory.

36. Colin Patterson, Address at the American Museum of Natural History, New York City, 5 November 1981, unpublished transcript. While Patterson's address raised much controversy, Phillip Johnson notes: "I

discussed evolution with Patterson for several hours in London in 1988. He did not retract any of the specific skeptical statements he has made, but he did say that he continues to accept 'evolution' as the only conceivable explanation for certain features of the natural world" (Phillip Johnson, *Darwin on Trial* [Downers Grove, IL: InterVarsity Press, 1993], 173).

3. Ape-Men Fiction, Fraud, and Fantasy

1. *Illustrated London News,* 24 June 1922, cited in Duane T. Gish, *Evolution: The Fossils Still Say No!* (El Cajon, CA: Institute for Creation Research, 1995), 327–28.

2. Adapted from Gish, 328.

3. Factual information adapted from Gish, 280–81. Marvin Lubenow explains that Dubois may well have concealed the discovery of the Wadjak skulls by declaring them only in "bureaucratic reports" that "were not intended for the public or for the scientific community" (Marvin L. Lubenow, *Bones of Contention: A Creationist Assessment of the Human Fossils* [Grand Rapids, MI: Baker Books, 1992], 104).

4. Sir Arthur Keith, *The Antiquity of Man,* rev. ed., 2 vols. (London: Williams and Norgate, Ltd., 1925), vol. 2, 440–41. As quoted in Lubenow, 102.

5. Lubenow, 115; see 113–19 for a good overview of the Selenka Expedition.

6. Michael D. Lemonick, "How Man Began," *Time,* 14 March 1994.

7. Henry M. Morris and Gary E. Parker, *What Is Creation Science?,* rev. ed. (El Cajon, CA: Master Books, 1987), 154.

8. Lubenow, 40–43. William R. Fix, *The Bone Peddlers: Selling Evolution* (New York: Macmillan, 1984), 12, 13.

9. Lubenow, 43.

10. Fix, 12.

11. Steve Jones, Robert Martin, and David Pilbeam, eds., *The Cambridge Encyclopedia of Human Evolution* (Cambridge: Cambridge University Press, 1992), 448.

12. Ibid. Lubenow, 42–43.

13. Peking man (originally classified as *Sinanthropus*) and Java man (*Pithecanthropus erectus*) have both been reclassified as belonging to the same species: *Homo erectus.*

14. Ian T. Taylor, *In the Minds of Men,* 3d ed. (Toronto: TFE Publishing, 1991), 235.

15. Ibid., 236.

16. Ibid., 237.

17. Ibid., 240; some of the most interesting facts for reconstructing the Peking man story are found on pages 234–41.

18. See Gish, 292–93 and Taylor, 238–40.

19. Evolutionary theory is so pervasive and has such a stranglehold on the scientific community that the pressure to conform all facts to the theory—no matter how contradictory to the theory the facts may be—is virtually insurmountable. In the field of paleoanthropology, in which evidence for evolution is particularly scant, the proclivity to rely on subjectivism is especially noticeable. This state of affairs goes a long way toward explaining the frequency of fantasy-filled fabrications in the history of hominid fossil studies. (See Phillip E. Johnson, *Darwin on Trial,* 2d ed. [Downers Grove, IL: InterVarsity Press, 1993], 81–87.)

20. *Ape Man: The Story of Human Evolution,* hosted by Walter Cronkite, Arts and Entertainment network, 4 September 1994.

21. *Modern People* 1 (18 April 1976): 11. As quoted in Gish, 309.

22. John Gribbon and Jeremy Cherfas, "Descent of

Man—or Ascent of Ape?" *New Scientist,* vol. 91 (1981), 592. As quoted in Gish, 311.

4. Chance

1. Jacques Monod, *Chance and Necessity* (New York: Vintage Books, 1972), 112–13. As quoted in John Ankerberg and John Weldon, *Darwin's Leap of Faith: Exposing the False Religion of Evolution* (Eugene, OR: Harvest House, 1998), 21.

2. R. C. Sproul, *Not a Chance: The Myth of Chance in Modern Science and Cosmology* (Grand Rapids, MI: Baker Books, 1994), 9.

3. Ibid., 3.

4. Chance as an ontological entity does not exist. So, when it is appealed to as an agency of cause, it is utterly impotent and meaningless. However, this sense of chance as a causal agency is what one gropes for in order to assert that universes appear out of nothing by chance. On the other hand, chance can quite usefully refer to formal mathematical probabilities, not at all signifying something that happens without a cause. In common parlance, when we say something has happened by chance, we don't mean that the event had no cause but that the actual cause is unknown to us. (See Sproul.)

5. Perhaps we should generously give evolutionists the benefit of the doubt at this point and assume that reference to chance here is not as an ontological causal agency (referring to the notion of uncaused effects) but in the formal sense of mathematical probabilities, such that an already existent material universe through time and natural processes alone might manifest life through some discernible causal pathway. Of course, as we will see, life cannot be produced in this way either.

6. James F. Coppedge, *Evolution: Possible or Impossible?* (Northridge, CA: Probability Research In Molecular Biology, 1993), 218.

7. Charles Darwin, *The Origin of Species by Means of Natural Selection.* As in Robert Maynard Hutchins, ed., *Great Books of the Western World*, vol. 49, *Darwin* (Chicago: Encyclopedia Britannica, 1952), 85.

8. Ibid. Of course, Darwin's lifework was intended to show that all biological organisms, with their attending "organs of extreme perfection and complication," were indeed formed through natural selection.

   Gordon Rattray Taylor points out further mind-boggling complications. Consider that in evolutionary mythology it is dogmatically asserted that snakes evolved from lizards despite the fact that the visual cells of lizards have no similarity to those of snakes. In addition, the eye appears suddenly in natural history, and even the earliest fishes have very sophisticated eyes. (Adapted from Gordon Rattray Taylor, *The Great Evolution Mystery* [New York: Harper & Row, Publishers, 1983], 101–2.)

9. Eye description adapted from G. Taylor, 101–2.

10. See Ibid., 98–103.

11. See Coppedge, 218–20. Michael Denton, *Evolution: A Theory in Crisis* (Bethesda, MD: Adler & Adler, 1985), 332–33.

12. Michael J. Behe, *Darwin's Black Box: The Biochemical Challenge to Evolution* (New York: The Free Press, 1996), 18. *Black box* is Behe's term for a device that does something but whose inner workings remain mysterious. For the average person, computers are a good example of a black box (p. 6).

13. Ibid., 22; see also 15–22.

14. In *Darwin's Black Box,* 18–21, Behe explains the biochemistry of vision:

> When light first strikes the retina a photon interacts with a molecule called 11-cis-retinal, which rearranges within picoseconds to trans-retinal. (A picosecond is about the time it takes light to travel the breadth of a single human hair.) The change in the shape of the retinal molecule forces a change in the shape of the

protein, rhodopsin, to which the retinal is tightly bound. The protein's metamorphosis alters its behavior. Now called metarhodopsin II, the protein sticks to another protein, called transducin. Before bumping into metarhodopsin II, transducin had tightly bound a small molecule called GDP. But when transducin interacts with metarhodopsin II, the GDP falls off, and a molecule called GTP binds to transducin. (GTP is closely related to, but critically different from, GDP.)

GTP-transducin-metarhodopsin II now binds to a protein called phosphodiesterase, located in the inner membrane of the cell. When attached to metarhodopsin II and its entourage, the phosphodiesterase acquires the chemical ability to "cut" a molecule called cGMP (a chemical relative of both GDP and GTP). Initially there are a lot of cGMP molecules in the cell, but the phosphodiesterase lowers its concentration, just as a pulled plug lowers the water level in a bathtub.

Another membrane protein that binds cGMP is called an ion channel. It acts as a gateway that regulates the number of sodium ions in the cell. Normally the ion channel allows sodium ions to flow into the cell, while a separate protein actively pumps them out again. The dual action of the ion channel and pump keeps the level of sodium ions in the cell within a narrow range. When the amount of cGMP is reduced because of cleavage by the phosphodiesterase, the ion channel closes, causing the cellular concentration of positively charged sodium ions to be reduced. This causes an imbalance of charge across the cell membrane that, finally, causes a current to be transmitted down the optic nerve to the brain. The result, when interpreted by the brain, is vision.

If the reactions mentioned above were the only ones that operated in the cell, the supply of 11-cis-retinal, cGMP, and sodium ions would quickly be depleted. Something has to turn off the proteins that were turned on and restore the cell to its original state. Several mechanisms do this. First, in the dark the ion channel (in addition to sodium ions) also lets calcium ions into the cell. The calcium is pumped back out by a different protein so that a constant calcium concentration is maintained. When cGMP levels fall, shutting down the ion channel, calcium ion concentration decreases, too. The phosphodiesterase enzyme, which destroys cGMP, slows down at lower calcium concentration. Second, a protein called guanylate cyclase begins to resynthesize cGMP when calcium levels start to fall. Third, while all of this is going on, metarhodopsin II is chemically modified by an enzyme called rhodopsin kinase. The modified rhodopsin then binds to a protein known as arrestin, which prevents the rhodopsin from activating more transducin. So the cell contains mechanisms to limit the amplified signal started by a single photon.

Trans-retinal eventually falls off of rhodopsin and must be reconverted to 11-cis-retinal and again bound by rhodopsin to get back to the starting point for another visual cycle. To accomplish this, trans-retinal is first chemically modified by an enzyme to trans-retinol—a form containing two more hydrogen atoms. A second enzyme then converts the molecule to 11-cis-retinol. Finally, a third enzyme removes the previously added hydrogen atoms to form 11-cis-retinal, a cycle is complete.

15. Phillip E. Johnson, *Defeating Darwinism by Opening Minds* (Downers Grove, IL: InterVarsity Press, 1997), 77.

16. Behe, 8–10.

17. Coppedge, 216, citing T. G. Taylor, "How an Eggshell Is Made," *Scientific American* (19 March 1970): 89–94.

18. Christopher Perrins, *Birds: Their Life, Their Ways, Their World* (Pleasantville, NY: Reader's Digest, 1979), 118–19.

19. *The Wonders of God's Creation: Human Life,* vol. 3, videotape (Chicago: Moody Institute of Science, 1993).

20. Ibid.

21. A. E. Wilder-Smith, *The Natural Sciences Know Nothing of Evolution* (Costa Mesa, CA: T.W.F.T. Publishers, 1981), 82.

22. A. E. Wilder-Smith, *The Origin of Life,* episode 3, videotape (Gilbert, AZ: Eden, 1983).

23. Coppedge, 50.

24. Ibid., 51.

25. Ibid., 53.

26. Ibid., 52. Coppedge explains the problem of trying to produce such a phrase by chance. The phrase, "the theory of evolution," contains twenty-three ordered letters and spaces. Thus, we need to randomly pick in an ordered sequence twenty-three specific objects out of a set of twenty-six letters of the alphabet and one "space." That means for the first "t" in our phrase there is a 1 in 27 chance of drawing it. Same with all the other letters in our phrase—each has a 1 in 27 chance of being drawn at any given time. But since we need the letters and spaces to come in a sequential order, we must multiply their separate probabilities. Since there are twenty-three letters and spaces to pick, and each has an individual probability of 1 in 27, we must multiply 27 by itself 23 times (i.e., 27^{23}). This means we would expect to succeed in spelling our phrase by chance only one time in over eight hundred million trillion trillion draws. Now, suppose we use a super computer to produce a billion draws per second. At this incredible rate we could expect to find, on average, only one successful spelling of our phrase in 26,000,000,000,000,000 years. This number of years is five million times as long as natural science estimates the earth to have existed. (Adapted from Coppedge, 52.)

If chance is so unproductive at producing such a simple phrase as "the theory of evolution," it is just inconceivable to think that chance could have produced something as organized and complex as a single cell, let alone the unfathomable, organized complexity of the human brain.

27. *The Wonders of God's Creation: Planet Earth,* vol. 1, videotape (Chicago: Moody Institute of Science, 1993).

28. Ibid.

29. Ibid.

30. Ibid. Scott M. Huse, *The Collapse of Evolution,* 2d ed. (Grand Rapids, MI: Baker Books, 1993), 71. Huse lists a number of additional demonstrations of design, including: (1) "Any appreciable change in the rate of rotation of the earth would make life impossible. For example, if the earth were to rotate at 1/10th its present rate, all plant life would either be burnt to a crisp during the day or frozen at night." (2) "The earth's axis is tilted 23.5 degrees from the perpendicular to the plane of its orbit. This tilting, combined with the earth's revolution around the sun, causes our seasons, which are absolutely essential for the raising of food supplies." (3) "The earth's atmosphere (ozone layer) serves as a protective shield from lethal solar ultraviolet radiation, which would otherwise destroy all life." (4) "The two primary constituents of the earth's atmosphere are nitrogen (78 percent) and oxygen (20

percent). This delicate and critical ratio is essential to all life forms." (5) "The earth's magnetic field provides important protection from harmful cosmic radiation" (pp. 71–72).

31. NOVA: *The Miracle of Life,* photographed by Lennart Nilsson, videotape (Boston: WGBH Educational Foundation, 1986, [Swedish Television Corp., 1982]); emphasis added. For a brief discussion, see Johnson, *Defeating Darwinism by Opening Minds,* 123.

32. Denton, 250.

33. Ibid.

34. Coppedge writes, "All known life on the earth consists largely of these giant molecules. 'The chemical basis of all life,' says the *Encyclopaedia Britannica,* 'is protein in a watery medium.'" Coppedge goes on to point out that the hemoglobin molecule, the most important protein molecule in blood, has 574 amino acid links and 10,000 atoms. In addition there are some 280,000,000 hemoglobin molecules per red blood cell. Insulin is the smallest molecule qualifying as a protein. Even it, however, has fifty-one amino acid links in two chains—one with twenty-one and the other with thirty amino acids. The two chains are joined together by sulfur bridges. The length of the average protein in the smallest known living thing is at least 400 amino acid links, containing more than 7,000 atoms. All the while every protein consists of a specific

exact sequence of these amino acid links (adapted from Coppedge, 98–102).

35. Coppedge, 110, 114.

36. Discussion adapted from Coppedge, 119–24. George Bernard Shaw used a similar argument to demonstrate that time and chance don't have a chance. Imagine a million monkeys typing on a million typewriters, twenty-four hours a day, one hundred words per minute, attempting to pound out a Shakespearean play. If each word contained four letters, the first word could be typed out by one of the monkeys in about twelve seconds. It would require about five days to get the first two words on one of the typewriters. To get the first four words would take one hundred billion years. The time it would take to type out Shakespeare's first scene, however, is beyond comprehension (adapted from Kenneth Boa and Larry Moody, *I'm Glad You Asked* [Wheaton, IL: Victor Books, 1982], 36).

    Evolutionists sometimes make the accusation that this type of argumentation does not correctly represent the evolutionary paradigm. The more sophisticated admit that the notion that chance alone is responsible for life is at best far-fetched. They suggest that rather than chance acting unilaterally, natural selection or some other unintelligent nonrandom mechanism was involved in the process. Perhaps beneficial molecular genetic changes are accumulated over time while natural selection weeds out harmful mutations (generally about one in 1,000 mutations are not harmful).

It should be noted that there is no evidence that suggests information in the genetic code is or can be increased in this manner. Nor are there any known physical laws that can be involved to account for the extremely high information content of genetic material. Furthermore, it is simply a logical fallacy to say that an accumulation of beneficial changes will produce an improved overall design. Finally, those capable of scaling the evolutionary language barrier realize that this is little more than using the phrase *natural selection* while pouring the meaning of intelligent design into the words. (See Nancy R. Pearcey, "DNA: The Message in the Message," *First Things* [June/July 1996]: 13–14. See also David Berlinski, "The Deniable Darwin," *Commentary* [June 1996].)

5. Empirical Science

1. For an excellent analysis of *Inherit the Wind,* see Phillip E. Johnson, *Defeating Darwinism by Opening Minds* (Downers Grove, IL: 1997), 24–36.

2. Discussion adapted from Henry M. Morris, *Men of Science, Men of God: Great Scientists Who Believed the Bible* (El Cajon, CA: Master Books, 1990).

3. Adapted from Ibid. cf. Fred Heeren, *Show Me God,* revised edition, (Wheeling, IL: Day Star, 1997), 334–63.

4. A. E. Wilder-Smith, *He Who Thinks Has to Believe* (Costa Mesa, CA: T.W.F.T. Publishers, 1981), 70. Wilder-Smith writes, "It is, of course, clear that Einstein did not claim to be a Christian. His convictions in metaphysical matters reached only to a firm belief in a Creator, which motivated Einstein's research in mathematics and physics" (p. 71).

5. Discussion adapted from Michael J. Behe, *Darwin's Black Box: The Biochemical Challenge to Evolution* (New York: Free Press, 1996), 197.

6. Heeren, 88.

7. Henry M. Morris, *Scientific Creationism,* public school edition (San Diego: C.L.P. Publishers, 1981), 19–20.

8. Kenneth Boa and Larry Moody, *I'm Glad You Asked* (Wheaton, IL: Victor Books, 1982), 38–39.

9. Carl Sagan, *Cosmos* (New York: Random House, 1980), 4. The PBS television series *Cosmos* was closely scripted after his book.

10. Albert Einstein, *Ideas and Opinions—The World as I See It* (New York: Bonanza Books, 1974), 40. As quoted in Heeren, 92. Heeren also writes that in 1917 Albert Einstein published a paper interpreting his theory of general relativity and made it conform to the cosmology of his day—the static universe theory. This claimed that

the universe is infinite in age, thus relieving the scientific community of having to deal with questions about the ultimate origin of the cosmos. Einstein was so convinced that these views were correct that he added what is now known as a cosmological "fudge factor" to his theory. Only after Einstein had seen evidence for an expanding universe by looking through Hubble's 100-inch telescope for himself did he formally renounce his cosmological "fudge factor," concerning which he later wrote was "the biggest blunder of my life." After this, Einstein wrote not only of the necessity of a beginning but also of his desire "to know how God created the world" (adapted from Heeren, xvi, xviii).

11. Heeren, 88–89.

12. The argument for God as an uncaused cause goes as follows: While modern science affirms the fact that the universe had a beginning, and thus is not eternal, the materialist is left with a dilemma. Either the universe sprang from nothing by chance, or something unbounded by time and greater than the universe caused it to come into being. In other words, either something came from nothing or something is eternal. Since we have previously demonstrated the absurdity of suggesting that something comes from nothing, we are forced to conclude that the universe was caused by something greater than itself—something or someone who ultimately is not dependent upon anything else in order to exist, something or someone who possesses the power of

being, intrinsically, and therefore is eternal. This eternal being we call God.

Some might say that if the universe needs a cause, then the cause of the universe, too, needs a cause. This sort of reasoning is what led thinkers such as Bertrand Russell to wonder who or what created God. But we can easily see where this reasoning leads. Either we would have an infinite regression of finite causes to account for the universe, which is self-contradictory and illogical (it doesn't answer the question of *source,* it merely makes the *effects* more numerous); or we must postulate that everything that ever exists comes from nothing by chance, which is also self-refuting. Or we can practice good science and encounter an uncaused cause. Russell's counter that if God doesn't need a cause, then neither does the universe, completely ignores that the nature of the universe is finite, changing, and therefore not eternal—so it is reasonable to ask, "Who made the universe?" God, however, doesn't need a cause because He is infinite and eternal. The universe must depend upon something that can account for its own existence, something eternal and, therefore, uncaused.

13. The first law of thermodynamics (conservation of energy) says that total energy, in all its forms, can neither be created nor destroyed, only shifted from one type to another. The second law of thermodynamics (law of entropy), in one of its several equivalent formulations, says that the amount of disorder in any isolated system cannot decrease with time. While the

total energy in the cosmos remains constant, the amount of energy available to do useful work is always getting smaller. (Discussion adapted from Robert M. Hazen and James Trefil, *Science Matters* [New York: Anchor Books, 1992], 24, 29–33.) The third law of thermodynamics says that the entropy of any pure crystal at absolute zero temperature is equal to zero—in a perfect crystal at absolute zero there is perfect order. (James E. Brady and Gerald E. Humiston, *General Chemistry: Principles and Structure,* 2d ed. [New York: Wyley, 1978], 315.)

14. *The New Encyclopedia Britannica,* Macropaedia, 15th ed., vol. 10 (Chicago: Encyclopedia Britannica, 1981), 415.

15. The energy equivalence of mass can be determined using Einstein's familiar formula, $E=mc^2$ (E is energy; m is mass; and c represents the speed of light).

16. Isaac Asimov, "In the Game of Energy and Thermodynamics You Can't Even Break Even," *Journal of Smithsonian Institute* (June 1970): 6. As quoted in Heeren, 128–29; also in Morris, *Scientific Creationism,* 21.

17. Heeren, 128.

18. Behe, 23–24.

19. Heeren, 129.

20. If one advances all of the information on one side of an issue and suppresses information on the other side, that by definition is not education but an attempt to seduce or trick people into believing a particular story—some may call it an attempt at brainwashing. While brainwashing is not actually possible, much of what passes for science is nothing more than philosophical pontification. Naturalism, which says that nature is all that exists, has become what Phillip Johnson calls "the established religious philosophy of America"—and anyone who attempts to buck this religious dogma faces tremendous pressure to conform. (For an excellent discussion, see Phillip E. Johnson, *Reason in the Balance* [Downers Grove, IL: InterVarsity Press, 1995].)

21. Arthur S. Eddington, *The Nature of the Physical World* (New York: Macmillan, 1930), 74. As quoted in Scott M. Huse, *The Collapse of Evolution* (Grand Rapids, MI: Baker Books, 1993), 77–78.

22. The DNA molecule carries instructions for building unfathomably complex living organisms. Thus, information entropy is perfectly relevant to the discussion of evolution on a genetic level. Information theory says that reduced entropy produced by random deviations is not equivalent to information—random processes do not produce information but rather distort information. Thus, random deviations in genetic material will not increase genetic information, which would be necessary for evolution to progress, let alone

produce DNA in the first place. (See A. E. Wilder-Smith, *The Natural Sciences Know Nothing of Evolution* [Costa Mesa, CA: T.W.F.T. Publishers, 1981], 69–73.)

23. Dr. Henry Morris demonstrates how the second law of thermodynamics (entropy) can be utilized in various contexts and defined in various ways, such as classical thermodynamics, statistical thermodynamics, and informational thermodynamics. In the first case, entropy is a measure of the unavailability of energy for further work. In the second case, entropy is a measure of the decreased order of a system's structure. And in the third case, it is a measure of lost or distorted information. In any case, what is being described is a downhill trend: Energy becomes unavailable, disorder increases, and information becomes garbled (adapted from *Scientific Creationism,* 38–40).

24. Myron Tribus and Edward C. McIrvine, "Energy and Information," *Scientific American* 224 (September 1971): 188. As quoted in Morris, *Scientific Creationism,* 39.

25. Raw energy is no better than zero energy at producing and sustaining life without something directing it—such as information and machinery. For example, a car sitting on an incline can progress downward but never upward without an engine and a conversion system to direct the raw energy of the gasoline in such a

way as to perform the work necessary to go uphill. Likewise, plants utilize the raw energy of the sun through the very complicated, information-dependent process of photosynthesis; without the information and machinery to perform photosynthesis, the sun's radiation would simply burn up the plant, which eventually does happen anyway. Raw energy without teleonomy is like a bull in a china shop, not a means of producing biological evolution. (For a helpful discussion see Morris, *Scientific Creationism,* 43–46.)

26. Willem J. J. Glashouwer and Paul S. Taylor, *The Origin of the Universe,* videotape (Mesa, AZ: Eden, 1983).

27. Morris, *Scientific Creationism,* 42.

28. Besides a preponderance of empirical evidence indicating that something does not come from nothing, the simple laws of logic require that nothing cannot produce anything—for nothing *is not* anything. It is a violation of the law of non-contradiction, which says that A is not non-A, to say that nothing can produce something. The thing said to be produced would have had to either create itself, or it would be an effect without a cause. If it created itself, it would have had to exist prior to its existence in order to do the creating, which means it must both exist and not exist in the same way and in the same respect, which is a violation of the law of non-contradiction. But if nothing caused it, it is said to be an effect without a cause. Not

only is this impossible by definition, since the definition of an effect involves a cause, it is impossible conceptually. Now, it is possible for something to exist without being an effect, but in order for something to exist and not be an effect it must be eternal (i.e., something that did not come into being, but always existed). God is such a being. But this fact in no way helps the case for an uncaused effect. If the laws of logic can be violated, then reason and communication are meaningless. (See R. C. Sproul, *Not a Chance: The Myth of Chance in Modern Science and Cosmology* [Grand Rapids, MI: Baker Books, 1994.])

29. Since the laws of thermodynamics remain unquestioned, we know the total amount of energy available to do work in the universe is not self-replenishing but is running out. (We can assume that the total available energy in the universe is finite since current cosmological models suggest this state of affairs.) Furthermore, we see that work is still being accomplished in the universe at this moment, which means we have not yet exhausted our finite supply of available energy. Since the universe in this respect is running downhill, and there is a finite supply of available energy, then the amount of time the universe has to exhaust its available energy is finite. But if the universe eternally existed, then an infinite amount of time has already passed. Infinite time would have consumed our universe's finite time in the infinite past—there would not be enough time left in the finite time available to our universe to last through

an infinite past. Since we are still here, the universe could not have had an eternal past. Therefore, the universe had a beginning and, thus, came into being.

6. Recapitulation

1. William R. Fix, *The Bone Peddlers* (New York: Macmillan, 1984), 285.

2. Stephen Jay Gould, *Ontogeny and Phylogeny* (Cambridge, MA: Bellknap Press, 1977), n.430.

3. Ibid., back cover. While admitting that the problems with Haeckel's recapitulation theory are myriad, Gould says he does not want to throw the baby out with the bath water:

   I am aware that I treat a subject currently unpopular. I do so, first of all, simply because it has fascinated me ever since the New York City public schools taught me Haeckel's doctrine, that ontogeny recapitulates phylogeny, fifty years after it had been abandoned by science. Yet I am not so detached a scholar that I would pursue it for the vanity of personal interest alone. I would not have spent some of the best years of a scientific career upon it, were I not convinced that it should be as important today as it has ever been.

   I am also not so courageous a scientist that I would have risked so much effort against a wall of truly universal opprobrium. But the chinks in the wall surfaced as soon as I probed. I have had the same, most curious experience more than twenty times: I tell a colleague that I am writing a book about parallels between ontogeny and phylogeny. He takes me aside, makes sure that no one is looking, checks for bugging devices, and admits in markedly lowered voice: "You know, just between you, me, and that wall, I think that there really is something to it after all." The clothing of disrepute is diaphanous before any good naturalist's experience. I feel like the honest little boy before the naked emperor.

I began this book as an indulgent, antiquarian exercise in personal interest. I hoped, at best, to retrieve from its current limbo the ancient subject of parallels between ontogeny and phylogeny. And a rescue it certainly deserves, for no discarded theme more clearly merits the old metaphor about throwing the baby out with the bath water. Haeckel's biogenetic law was so extreme, and its collapse so spectacular, that the entire subject became taboo. . . .

But I soon decided that the subject needs no apology. Properly restructured, it stands as a central theme in evolutionary biology because it illuminates two issues of great contemporary importance: the evolution of ecological strategies and the biology of regulation. . . .

That some relationship [between ontogeny and phylogeny] exists cannot be denied. Evolutionary changes must be expressed in ontogeny, and phyletic information must therefore reside in the development of individuals. This, in itself, is obvious and unenlightening. This book emphasizes the importance of one kind of relationship—the *changes in developmental timing* that produce *parallels* between the stages of ontogeny and phylogeny. (Ibid., 1–2; emphasis in original.)

4. Ian T. Taylor, *In the Minds of Men,* 3d ed. (Toronto: TFE Publishing, 1991), 274.

5. Ibid., 276. Ian Taylor writes:

   Haeckel stated that the ova and embryos of different vertebrate animals and man are, at certain periods of their development, all perfectly alike, indicating their supposed common origin. Haeckel produced the well-known illustration showing embryos at several stages of development. In this he had to play fast and loose with the facts by altering several drawings in order to make them appear more alike and conform to the theory. . . . In a catalog of errors, His (1874) showed that Haeckel had used two drawings of embryos, one taken from Bischoff (1845) and the other from Ecker (1851–59), and he had added 3–5 mm to the head of Bischoff's dog embryo, taken 2 mm off the head of Ecker's human embryo, reduced the size of the eye 5 mm, and doubled the length of the posterior. [See illustration, p. 94.]

6. Walt Brown, *In the Beginning: Compelling Evidence for Creation and the Flood,* 6th ed. (Phoenix: Center for Scientific Creation, 1995), 45.

7. Assmusth and Hull, *Haeckel's Frauds and Forgeries*

(India: Bombay Press, 1911). Cited in Luther D. Sunderland, *Darwin's Enigma,* 4th ed. rev. (Santee, CA: Master Books, 1988), 120.

8. Henry M. Morris, *Scientific Creationism,* public school edition (San Diego: C.L.P. Publishers, 1981), 77.

9. Ibid. See pp. 75–78 for more detail concerning alleged evolutionary vestiges and recapitulations.

10. Carl Sagan, *The Dragons of Eden* (New York: Random House, 1977), 57–58.

11. The throat (or pharyngeal) grooves and pouches, falsely called "gill slits," are *not* mistakes in human development. They develop into absolutely essential parts of human anatomy. The middle ear canals come from the second pouches, and the parathyroid and thymus glands come from the third and the fourth. Without a thymus, we would lose half our immune systems. Without the parathyroids, we would be unable to regulate calcium balance and could not even survive. Another pouch, thought to be vestigial by evolutionists until just recently, becomes a gland that assists in calcium balance. (Henry M. Morris and Gary E. Parker, *What Is Creation Science?,* rev. ed. [El Cajon, CA: Master Books, 1987], 64.)

12. Sagan, *The Dragons of Eden,* 197.

13. Ibid.

14. Ambassador Curtin Winsor, Jr., "Letter to the Editor" in *National Review* (2 June 1989): 8.

15. Ibid.

16. Elie Schneour, "Life Doesn't Begin, It Continues," *Los Angeles Times* (29 January 1985): part v. As quoted in Morris, *The Long War Against God,* 138.

17. *The Human Life Bill: Hearings on S. 158 Before the Subcommittee on Separation of Powers of the Senate Judiciary Committee,* 97th Congress, 1st Session (1981). As quoted in Norman L. Geisler, *Christian Ethics: Options and Issues* (Grand Rapids, MI: Baker Books, 1989), 149; cited in Francis J. Beckwith, *Politically Correct Death: Answering the Arguments for Abortion Rights* (Grand Rapids, MI: Baker, 1993), 42.

18. *The Human Life Bill—S. 158,* Report, 9. As quoted in Francis J. Beckwith, *Politically Correct Death: Answering the Arguments for Abortion Rights* (Grand Rapids, MI: Baker Books, 1993), 42.

19. Morris and Parker, 67. See Stephen Jay Gould, "Dr. Down's Syndrome," *Natural History* (April 1980): 142–48.

20. Henry M. Morris, *Creation and the Modern Christian* (El Cajon, CA: Master Books, 1985), 72.

21. Ibid.

22. Henry M. Morris, *Science and the Bible,* rev. ed. (Chicago: Moody Press, 1986), 50.

23. As quoted in *Natural History* (April 1980): 129, cited in Morris, *Creation and the Modern Christian,* 72.

24. Morris, *The Long War Against God,* 139.

25. Stephen Jay Gould, "Dr. Down's Syndrome," *Natural History* (April 1980): 144. As quoted in Morris, *Creation and the Modern Christian,* 71.

26. Jacques Monod, "The Secret of Life," interview with Laurie John, Australian Broadcasting Co., 10 June 1976. As quoted in Morris, *The Long War Against God,* 58.

27. Furthermore, the entire evolutionary program is driven by *naturalistic* assumptions, not theistic ones. There is no point in invoking God to account for a process in which His presence is, *a priori,* not needed.

28. G. Richard Bozarth, "The Meaning of Evolution," *American Atheist* (February 1978): 19.

29. David Berlinski, *A Firing Line Debate,* with William F. Buckley, Jr. (Public Broadcasting System, 19 December 1997).

Epilogue

1. Romans 3:9–20 shows that all people, Jew or Gentile, atheist or pagan, are separated from God by their sin. Romans 3:23 uses the same theme but refers specifically to sinners who, by the grace of God, are not condemned but redeemed (see verse 24).

2. Some argue that this verse applies only to believers. However, I'm convinced that anyone who takes the time to examine Revelation 3 in context would agree that the church of Laodicea was spiritually pitiful, poor, blind, and naked. While in context the passage is applied to the Laodicean church corporately, it most certainly has personal application as well.

Appendix A. Death Moves

1. Jim McLean, *The Eight-Step Swing* (New York: HarperCollins, 1994), xv.

2. James Moore, *The Darwin Legend* (Grand Rapids, MI: Baker Books, 1994), back cover.

3. John Arnott, "The Love of God," sermon delivered at Vineyard Christian Fellowship, Mission Viejo, California, 17 July 1995, audio tape #621, transcript.

4. Moore, back cover.

5. Ibid., 12.

Appendix C. The Greatest FEAT in the Annals of Recorded History

1. Simon Greenleaf, *The Testimony of the Evangelists* (Grand Rapids, MI: Baker Books, 1984 [orig. New York: Cockcroft, 1874]), 29–30.

Appendix D. Annihilating Abortion Arguments

1. Francis A. Schaeffer and C. Everett Koop, *Whatever Happened to the Human Race?* in *The Complete Works of Francis A. Schaeffer: A Christian Worldview,* 5 vols. (Wheaton, IL: Crossway, 1982), 5:293.

2. Quoted in *Policy Review* (Spring 1985): 15. This, along with the following four quotes, can be found in Francis J. Beckwith, *Politically Correct Death: Answering the Arguments for Abortion Rights* (Grand Rapids, MI: Baker Books, 1993), 174.

3. Debate with Francis J. Beckwith on the campus of the University of Nevada, Las Vegas, December 1989.

4. Quoted in Robert Marshall and Charles Donovan, *Blessed Are the Barren: The Social Policy of Planned Parenthood* (San Francisco: Ignatius, 1991), 182.

5. Margaret Sanger, *Women and the New Race* (New York: Brentano's, 1920), 63.

6. *AMA Prism* (May 1993): 2.

7. See James C. Dobson, *Focus on the Family Newsletter,* July 1993.

8. Ibid.

9. Ibid., 2.

10. *The Human Life Bill—S. 158, Report together with Additional and Minority Views to the Committee on the Judiciary, United States Senate, made by its Subcommittee on Separation of Powers,* 97th Congress, 1st Session (1981), 11. As quoted in Beckwith, 43.

11. *The Human Life Bill: Hearings on S. 158 before the Subcommittee on Separation of Powers of the Senate Judiciary Committee,* 97th Congress, 1st Session (1981). As quoted in Norman L. Geisler, *Christian Ethics: Options and Issues* (Grand Rapids, MI: Baker Books, 1989), 149; cited in Beckwith, 42.

12. *The Human Life Bill—S. 158, Report,* 9. As quoted in Beckwith, 42.

13. See Beckwith, 88.

14. Quoted in Michael Ross, "Senate Bans Use of Force Against Abortion Clinics," *Los Angeles Times,* 17 November 1993, A1.

15. Ibid., A1, A22.

16. Karl Marx, *Critique of the Hegelian Philosophy of Right* (1843–1844; introduction). As quoted in John Bartlett, *Bartlett's Familiar Quotations* 16th ed. (Boston: Little, Brown, 1982), 481.

17. Charles R. Hayman and Charlene Lanza, "Sexual Assault in Women and Girls," *American Journal of Obstetrics and Gynecology* 109 (1971): 480–86; cited in Beckwith, 241 n. 69.

18. Beckwith, 69.

19. Bernard Nathanson, *Aborting America* (New York: Doubleday, 1979), 193. As quoted in Beckwith, 55.

20. Ibid.

21. Beckwith, 43.

Appendix E. Seed's Preposterous Plan

1. *USA Today,* 7 January 1998, Nation (www.USA.com).

2. Linda Chavez, "Cloning and Abortion: Two Sides of a Coin," *Orange County Register,* 15 January 1998, Metro section.

3. Ibid., which references Supreme Court decision,

Griswold v. Connecticut, 1964.

4. Ibid., which references Supreme Court decision, *Roe v. Wade,* 1973.

5. Ibid.

6. Ray Bohlin, "Can Humans Be Cloned Like Sheep?" (Richardson, TX: Probe Ministries, 1997), 2 (www.probe.org). Bohlin notes that out of 277 initial cell fusions, 29 began growing as embryos in vitro. All 29 embryos were implanted into 13 receptive ewes, yet only one lamb was born.

7. It is important to note that while it is theoretically possible to clone a human body, the human spirit can never be cloned. As with twins, the clone may share a genetic blueprint with the body from which it is derived but its soul or spirit would nevertheless remain distinct.

8. Joseph Schuman, "European Nations Sign Ban on Human Cloning," *Orange County Register,* 13 January 1998, front page.

9. Quoted in Francis J. Beckwith, *Politically Correct Death: Answering the Arguments for Abortion Rights* (Grand Rapids, MI: Baker Books, 1993), 42. Many other leading medical experts echo LeJeune's words. These experts include Dr. Micheline Matthews-Roth, a principal research associate in the department of

Medicine at Harvard (Beckwith, 43), and Dr. Hymie Gordon, professor of medical genetics and a physician at the Mayo Clinic (ibid., 42). While pro-abortionists try to separate "human life" from "personhood," this distinction is merely a semantic deception. Personhood is an essential attribute of human nature; if one is a human life, he or she is naturally also a human person.

10. In essence, cloning is a form of asexual reproduction.

Bibliographical Data for Epigraphs

Before You Begin (p. 1)
—Psalm 19: 1-4.

Charting the Course (p. 7)
—F. Darwin ed., *The Life and Letters of Charles Darwin,* vol. 1 (London: John Murray, 1888), 307. As quoted in Michael Denton, *Evolution: A Theory in Crisis* (Bethesda, MD: Adler & Adler, 1986), 25.
—Charles Darwin, *The Origin of Species By Means of Natural Selection,* chap. VI, "Difficulties of the Theory," in Robert Maynard Hutchins, ed., *Great Books of the Western World,* vol. 49, Darwin (Chicago: Encyclopaedia Britannica, 1952), 85.

Chapter 1: Truth or Consequences (p. 15)
—As quoted in Henry M. Morris, *The Long War Against God*

(Grand Rapids, MI: Baker Book House, 1989), 22.
—Genesis 1:11, 24-25.

Chapter 2: Fossil Follies (p. 31)
—Charles Darwin, *The Origin of Species By Means of Natural Selection,* chap. X, "On the Imperfection of the Geological Record," in Robert Maynard Hutchins, ed., *Great Books of the Western World,* vol. 49, Darwin (Chicago: Encyclopaedia Britannica, 1952), 152.
—David M. Raup, "Conflicts Between Darwin and Paleontology," *Bulletin, Field Museum of Natural History,* January 1979, 22 (emphasis added). As quoted in Paul S. Taylor, *The Illustrated Origins Answer Book,* fourth edition (Mesa, AZ: Eden Productions, 1993), 108.

Chapter 3: Ape-Men Fiction, Fraud, and Fantasy (p. 47)
—*Ape Man: The Story of Human Evolution,* hosted by Walter Cronkite, Arts and Entertainment network, 4 September 1994.
—Genesis 1:27.

Chapter 4: Chance (p. 59)
—Jacques Monod, *Chance and Necessity: An Essay on the Natural Philosophy of Modern Biology* (New York: Alfred A. Knopf, 1971), 112-113.
—David Hume, *An Enquiry Concerning Human Understanding,* Section VIII, part I. Quoted in R. C. Sproul, *Not A Chance: The Myth of Chance in Modern Science and Cosmology* (Grand Rapids, MI: Baker Books, 1994), 193.

Chapter 5: Empirical Science (p. 75)

—Richard Dawkins, "The Emptiness of Theology," *Free Inquiry,* May 1998, 6.

—Wernher von Braun, *Creation: Nature's Designs and Designer* (Mountain View, CA: Pacific Press Publishing Association, 1971), 6. As quoted in John Ankerberg and John Weldon, *Darwin's Leap of Faith: Exposing the False Religion of Evolution* (Eugene, OR: Harvest House Publishers, 1998), 129-130.

Chapter 6: Recapitulation (p. 91)

—Elie A. Schneour, "Life Doesn't Begin, It Continues," *Los Angeles Times,* 29 January 1989, Part V. As quoted in Henry M. Morris, *The Long War Against God* (Grand Rapids, MI: Baker Book House, 1989), 138.

—Stephen Jay Gould, "Dr. Down's Syndrome," *Natural History,* April 1980, 144. Cited in Henry M. Morris, *Creation and the Modern Christian,* (El Cajon, CA: Master Book Publishers, 1985), 71.

Epilogue (p. 105)

—Romans 1:20.

Bibliography

Creation/Evolution

Books

Ankerberg, John, and John Weldon. *Darwin's Leap of Faith.* Eugene, OR: Harvest House, 1998.

Behe, Michael J. *Darwin's Black Box: The Biochemical Challenge to Evolution.* New York: Free Press, 1996.

Bird, W. R. *The Origin of Species Revisited,* 2 vols. New York: Philosophical Library, 1989.

Boa, Kenneth, and Larry Moody. *I'm Glad You Asked.* Wheaton, IL: Victor Books, 1982.

Brady, James E., and Gerald E. Humiston. *General Chemistry: Principles and Structure.* 2d ed. New York: Wyley, 1978.

Brown, Walt. *In the Beginning: Compelling Evidence for Creation and the Flood.* 6th ed. Phoenix: Center for Scientific Creation, 1995.

Coppedge, James F. *Evolution: Possible or Impossible?* Northridge, CA: Probability Research in Molecular Biology, 1993.

Darwin, Charles. *The Descent of Man.* In *Great Books of the Western World,* vol. 49, *Darwin,* edited by Robert Maynard Hutchins. Chicago: Encyclopedia Britannica, 1952.

———.*The Life and Letters of Charles Darwin,* vol. 1 (London: John Murray, 1888).

———. *The Origin of Species by Means of Natural Selection.* In *Great Books of the Western World,* vol. 49, *Darwin,* edited by Robert Maynard Hutchins. Chicago: Encyclopedia Britannica, 1952.

Davis, Percival, and Dean H. Kenyon. *Of Pandas and People: The Central Question of Biological Origins.* Dallas, TX: Haughton, 1989.

Denton, Michael. *Evolution: A Theory in Crisis.* Bethesda, MD: Adler & Adler, 1986.

du Nouy, Pierre Lecomte. *Human Destiny.* New York: Longmans, Green, 1947.

Fix, William R. *The Bone Peddlers: Selling Evolution.* New York: Macmillan, 1984.

Gish, Duane T. *Evolution: The Fossils Still Say No!* El Cajon, CA: Institute for Creation Research, 1995.

Gould, Stephen Jay. *Ontogeny and Phylogeny.* Cambridge, MA: Bellknap Press, 1977.

Hazen, Robert M., and James Trefil. *Science Matters.* New York: Anchor Books, 1992.

Heeren, Fred. *Show Me God*. Rev. ed. Wheeling, IL: Day Star, 1997.

Hume, David. *An Enquiry Concerning Human Understanding*. Indianapolis, IN: Hackett, 1993.

Huse, Scott M. *The Collapse of Evolution*, 2d ed. Grand Rapids, MI: Baker Books, 1993.

Huxley, Aldous. *Ends and Means*. London: Chatto & Windus, 1938.

Johnson, Phillip E. *Darwin on Trial*. Downers Grove, IL: InterVarsity Press, 1993.

———. *Defeating Darwinism by Opening Minds*. Downers Grove, IL: InterVarsity Press, 1997.

———. *Reason in the Balance*. Downers Grove, IL: InterVarsity Press, 1995.

Jones, Steve, Robert Martin, and David Pilbeam, eds. *The Cambridge Encyclopedia of Human Evolution*. Cambridge: Cambridge University Press, 1992.

Lubenow, Marvin L. *Bones of Contention: A Creationist Assessment of the Human Fossils*. Grand Rapids, MI: Baker Books, 1992.

Marshall, Robert, and Charles Donovan. *Blessed Are the Barren: The Social Policy of Planned Parenthood*. San Francisco: Ignatius, 1991.

McLean, Jim. *The Eight-Step Swing*. New York: Harper-Collins, 1994.

Monod, Jacques. *Chance and Necessity*. New York: Vintage Books, 1972.

Moore, James. *The Darwin Legend*. Grand Rapids, MI: Baker Books, 1994.

Morris, Henry M. *Creation and the Modern Christian*. El Cajon, CA: Master Books, 1985.

———. *The Long War Against God.* Grand Rapids, MI: Baker Books, 1989.

———. *Men of Science Men of God: Great Scientists Who Believed the Bible.* El Cajon, CA: Master Books, 1990.

———. *Science and the Bible.* Rev. ed. Chicago: Moody Press, 1986.

———. *Scientific Creationism.* Public school ed. San Diego: C.L.P. Publishers, 1981.

———. *That Their Words May Be Used Against Them.* El Cajon, CA: Institute for Creation Research, 1997.

Morris, Henry M., and Gary E. Parker. *What Is Creation Science?* Rev. ed. El Cajon, CA: Master Books, 1987.

The New Encyclopedia Britannica. Micropaedia, 15th ed. Vol. 1. Chicago: Encyclopedia Britannica, 1981.

The New Encyclopedia Britannica. Macropaedia, 15th ed. Vol. 10. Chicago: Encyclopedia Britannica, 1981.

Perrins, Christopher. *Birds: Their Life, Their Ways, Their World.* Pleasantville, NY: Reader's Digest, 1979.

Rehwinkel, Alfred. M. *The Wonders of Creation.* Grand Rapids, MI: Baker Books, 1974.

Ross, Hugh. *The Creator and the Cosmos.* Colorado Springs, CO: NavPress, 1993.

———. *The Fingerprint of God.* 2d ed. Orange, CA: Promise, 1991.

Sagan, Carl. *Cosmos.* New York: Random House, 1980.

———. *The Dragons of Eden.* New York: Random House, 1977.

Saint, Phil. *Fossils That Speak Out.* Greensboro, NC: Saint Ministries, 1985.

Sproul, R. C. *Not a Chance: The Myth of Chance in Modern Science and Cosmology.* Grand Rapids, MI: Baker Books, 1994.

Starr, Cecie. *Biology: Concepts and Applications.* 3d ed. Belmont, CA: Wadsworth, 1997.

Sunderland, Luther D. *Darwin's Enigma.* 4th ed. Santee, CA: Master Books, 1988.

Taylor, Gordon Rattray. *The Great Evolution Mystery.* New York: Harper & Row, 1983.

Taylor, Ian T. *In the Minds of Men.* 3d ed. Toronto: TFE Publishing, 1991.

Taylor, Paul S. *The Illustrated Origins Answer Book.* 4th ed. Mesa, AZ: Eden, 1993.

Thaxton, Charles B., Walter L. Bradley, and Roger L. Olsen. *The Mystery of Life's Origin: Reassessing Current Theories.* New York: Philosophical Library, 1994.

Van Till, Howard J., Davis Young, and Clarence Menninga. *Science Held Hostage: What's Wrong with Creation Science AND Evolutionism.* Downers Grove, IL: InterVarsity Press, 1988.

Von Braun, Wernher. *Creation: Nature's Designs and Designer.* Mountain View, CA: Pacific Press Publishing Assoc., 1971.

Wilder-Smith, A. E. *The Creation of Life: A Cybernetic Approach to Evolution.* Costa Mesa, CA: T.W.F.T. Publishers, 1970, 1988.

———. *He Who Thinks Has to Believe.* Costa Mesa, CA: T.W.F.T. Publishers, 1981.

———. *The Natural Sciences Know Nothing of Evolution.* Costa Mesa, CA: T.W.F.T. Publishers, 1981.

———. *The Scientific Alternative to Neo-Darwinian Evolutionary Theory: Information Sources and Structures.* Costa Mesa, CA: T.W.F.T. Publishers, 1987.

Articles

Berlinski, David. "The Deniable Darwin." *Commentary* (June 1996): reprint.

Bozarth, G. Richard. "The Meaning of Evolution." *American Atheist* (February 1978): 19, 33.

Colby, Chris. "Introduction to Evolutionary Biology," version 2. The Talk.Origins Archive. (7 January 1996, www.talk-origins.org/faqs/faq-intro-to-biology.html).

Dawkins, Richard. "The Emptiness of Theology." *Free Inquiry* (May 1998): 6.

Dembski, Bill. "The Explanatory Filter: A Three-Part Filter for Understanding How to Separate and Identify Cause from Intelligent Design." An excerpt from a paper presented at the 1996 Mere Creation Conference. (www.origins.org/real/ri9602/dembski.html)

Dobson, James. "Family News from Dr. James Dobson." *Focus on the Family Newsletter* (February 1994).

Gould, Stephen Jay. "Dr. Down's Syndrome." *Natural History* (April 1980): 142–48.

———. "The Five Kingdoms." *Natural History* (June 1976): 30–37.

———. "Hooking Leviathon by Its Past." *Natural History* (May 1994): 8–15.

———. "Evolution's Erratic Pace." *Natural History* (May 1977): 12–16.

———. "Piltdown Revisited." *Natural History* (March 1979): 86–97.

———. "The Return of Hopeful Monsters." *Natural History* (June 1977): 22–30.

Isaak, Mark. "Five Major Misconceptions about Evolution." The Talk.Origins Archive. (3 January 1995, www.talkorigins.org/faqs/faq-misconceptions.html).

Lemonick, Michael D. "How Man Began." *Time* (14 March 1994): reprint.

Patterson, Colin. Address at the American Museum of Natural History, New York City, (5 November 1981, unpublished transcript).

Pearcey, Nancy R. "DNA: The Message in the Message." *First Things* (June/July 1996): 13–14. (www.origins.org/ftissues/ft9606/pearcy.html).

Raup, David M. "Conflicts between Darwin and Paleontology." *Bulletin, Field Museum of Natural History* (January 1979): 22, 25.

Schneour, Elie A. "Life Doesn't Begin, It Continues." *Los Angeles Times* (29 January 1989): Part V.

Winsor, Curtin Jr. "Letter to the Editor." *National Review* (2 June 1989): 8.

Audiotapes

Arnott, John. "The Love of God." Mission Viejo, CA: Vineyard Christian Fellowship, 17 July 1995 (#621, transcript).

Videotapes

Ape Man: The Story of Human Evolution, hosted by Walter Cronkite. Arts and Entertainment, 4 September 1994.

A Firing Line Debate, with William F. Buckley, Jr. PBS, 21 December 1997.

Ham, Kenneth, and Gary Parker. *Understanding Genesis.* 10 tapes. El Cajon, CA: Creation Life, n.d.

NOVA: The Miracle of Life, photographed by Lennart Nilsson. Boston: WGBH Educational Foundation, 1986 (Swedish Television Corp., 1982).

The Origin of the Universe. featuring A. E. Wilder-Smith. 6 parts. Mesa, AZ: Eden Films and Standard Media, 1983.

The Wonders of God's Creation. 3 tapes. Chicago: Moody Institute of Science, 1993.

Websites

Origins. (www.origins.org)

Talk.Origins Archive. (www.talkorigins.org)

The Alan Guttmacher Institute. (www.agi-usa.org)

The Centers for Disease Control and Prevention. (www.cdc.gov)

Department of Veterans Affairs. (www.va.gov)

Probe Ministries. (www.probe.org)

Abortion

Books

Beckwith, Francis J. *Politically Correct Death: Answering Arguments for Abortion Rights.* Grand Rapids, MI: Baker Books, 1993.

Geisler, Norman L. *Christian Ethics: Options and Issues.* Grand Rapids, MI: Baker Books, 1989.

Kreeft, Peter. *The Unaborted Socrates.* Downers Grove, IL: InterVarsity Press, 1983.

Marshall, Robert, and Charles Donovan. *Blessed Are the Barren: The Social Policy of Planned Parenthood.* San Francisco: Ignatius, 1991.

Sanger, Margaret. *Women and the New Race.* New York: Brentano's, 1920.

Schaeffer, Francis A., and C. Everett Koop. *Whatever*

Happened to the Human Race? In *The Complete Works of Francis A. Schaeffer: A Christian Worldview.* 5 vols. Wheaton, IL: Crossway, 1982.

Articles

AMA Prism (May 1993): 2.

Beckwith, Francis J. "Answering the Arguments for Abortion Rights, Part One: The Appeal to Pity." *Christian Research Journal* (Fall 1990).

———. "Answering the Arguments for Abortion Rights, Part Two: Arguments from Pity, Tolerance, and Ad Hominem." *Christian Research Journal* (Winter 1991).

Dobson, James A. *Focus on the Family Newsletter* (July 1993).

Ross, Michael. "Senate Bans Use of Force Against Abortion Clinics." *Los Angeles Times,* 17 November 1993, A1.

Audiotape

Debate with Francis J. Beckwith on the campus of the University of Nevada, Las Vegas, December 1989.

Videotape

Cunningham, Greg (narrator). *Hard Truth.* Cleveland: American Portrait Films, 1991.

Cloning

Articles

Bohlin, Ray. "Can Humans Be Cloned Like Sheep?" Richardson, TX: Probe Ministries, 1997 (www.probe.org).

Chavez, Linda. "Cloning and Abortion: Two Sides of a Coin." *Orange County Register,* 15 January 1998, Metro section.

Schuman, Joseph. "European Nations Sign Ban on Human Cloning." *Orange County Register,* 13 January 1998, front page.

USA Today, 7 January 1998, Nation (www.USA.com).

Scripture Index

Subject Index

B

C

E

F

G

About the Author

Hank Hanegraaff answers questions live as host of the "Bible Answer Man" broadcast, heard daily throughout the United States and Canada. He is president of the world-renowned Christian Research Institute headquartered in southern California and author of the best-selling Gold Medallion winner *Christianity in Crisis* as well as the award winning best-seller *Counterfeit Revival.*

As author of Memory Dynamices, Hank has developed memorable tools to prepare believers to effectively communicate 1) *what* they believe, 2) *why* they believe it, and 3) *where* cults deviate from historic Christianity. He has also developed fun and easy techniques for memorizing Scripture quickly and retaining it forever.

Hank is a popular conference speaker for churches and conferences worldwide. He resides in Southern California with his wife, Kathy and eight children: Michelle, Katie,

David, John, Mark, Hank Jr., Christina, Paul, and Faith.

For further information on Memory and Personal Witness Training materials address your request to:

Hank Hanegraaff
Box 80250
Rancho Santa Margarita, CA 92688-0250
or call (949) 589-1504

The FACE That Demonstrates the Farce of Evolution on Tape

Hank Hanegraaff is passionate about the issue of origins. In the audio book version of *The FACE* you can experience that passion for yourself. Your ability to absorb this already memorable material will be greatly enhanced as it enters your mind through both the *ear* gate and the *eye* gate.